Basic Gambling Mathematics

Basic Gambling Mathematics: The Numbers Behind the Neon, Second Edition explains the mathematics involved in analyzing games of chance, including casino games, horse racing and other sports, and lotteries. The book helps readers understand the mathematical reasons why some gambling games are better for the player than others. It is also suitable as a textbook for an introductory course on probability.

Along with discussing the mathematics of well-known casino games, the author examines game variations that have been proposed or used in actual casinos. Numerous examples illustrate the mathematical ideas in a range of casino games while end-of-chapter exercises go beyond routine calculations to give readers hands-on experience with casino-related computations.

New to the Second Edition

- Thorough revision of content throughout, including new sections on the birthday problem (for informal gamblers) and the Monty Hall problem, as well as an abundance of fresh material on sports gambling
- Brand new exercises and problems
- A more accessible level of mathematical complexity, to appeal to a wider audience.

Mark Bollman is Professor of Mathematics and chair of the Department of Mathematics & Computer Science at Albion College in Albion, Michigan, and has taught 118 different courses in his career. Among these courses is "Mathematics of the Gaming Industry," where mathematics majors carefully study the math behind games of chance and travel to Las Vegas, Nevada, in order to compare theory and practice. He has also taken those ideas into Albion's Honors Program in "Great Issues in Humanities: Perspectives on Gambling," which considers gambling from literary, philosophical, and historical points of view as well as mathematically. Mark has also authored *Mathematics of Keno and Lotteries, Mathematics of Casino Carnival Games,* and *Mathematics of The Big Four Casino Table Games: Blackjack, Baccarat, Craps, & Roulette.*

AK Peters/CRC Recreational Mathematics Series

Series Editors

Robert Fathauer
Snezana Lawrence
Jun Mitani
Colm Mulcahy
Peter Winkler
Carolyn Yackel

Mathematical Puzzles
Peter Winkler

X Marks the Spot
The Lost Inheritance of Mathematics
Richard Garfinkle, David Garfinkle

Luck, Logic, and White Lies
The Mathematics of Games, Second Edition
Jörg Bewersdorff

Mathematics of The Big Four Casino Table Games
Blackjack, Baccarat, Craps, & Roulette
Mark Bollman

Star Origami
The Starrygami™ Galaxy of Modular Origami Stars, Rings and Wreaths
Tung Ken Lam

Mathematical Recreations from the Tournament of the Towns
Andy Liu, Peter Taylor

The Baseball Mysteries
Challenging Puzzles for Logical Detectives
Jerry Butters, Jim Henle

Mathematical Conundrums
Barry R. Clarke

Lateral Solutions to Mathematical Problems
Des MacHale

Basic Gambling Mathematics
The Numbers Behind the Neon, Second Edition
Mark Bollman

For more information about this series please visit: https://www.routledge.com/AK-PetersCRC-Recreational-Mathematics-Series/book- series/RECMATH?pd=published,forthcoming&pg=2&pp=12&so=pub&view=list

Basic Gambling Mathematics
The Numbers Behind the Neon
Second Edition

Mark Bollman
Department of Mathematics
Albion College, Albion, MI, USA

CRC Press
Taylor & Francis Group
Boca Raton London New York

CRC Press is an imprint of the
Taylor & Francis Group, an **informa** business

AN A K PETERS BOOK

ISBN: 978-1-032-41461-4 (hbk)
ISBN: 978-1-032-41460-7 (pbk)
ISBN: 978-1-003-35818-3 (ebk)

DOI: 10.1201/9781003358183

Typeset in Latin Modern font
by KnowledgeWorks Global Ltd.

Publisher's note: This book has been prepared from camera-ready copy provided by the author.

*Dedicated to the memory of
Harry O. Zimmerman II and Linda D. Morris,
who are largely responsible for my interest in
what happens in Las Vegas.*

Contents

Preface

This book grew out of several years of teaching about gambling in a variety of contexts at Albion College beginning in 2002. For several years, I taught a first-year seminar called "Chance," which I came to describe as "probability and statistics for the educated citizen" as distinguished from a formula-heavy approach to elementary statistics. I also focused more on probability than statistics in Chance. Part of probability is gambling, of course, and so over the years, the course evolved to include more casino examples in class, whether by simulation or actual in-class game play. The course included a field trip to the Soaring Eagle Casino in Mount Pleasant, Michigan, late in the semester after all of the students had turned 18. This provided students with a fine opportunity to combine theory with practice and see for themselves how the laws of probability worked, in a way that no classroom activity could mimic.

Later on, I expanded the gambling material into a course called Great Issues in Humanities: Perspectives on Gambling, in Albion's honors program. The course combined mathematics from Chance (for mathematics, in the words of one of my colleagues, is the first of the humanities) with other readings from literature, philosophy, and history to provide a well-rounded view of a subject that is not becoming less important in America.

Throughout my years teaching about gambling, I struggled to find a good probability textbook that covered the topics germane to my course without a lot of material that was not related to gambling. *Basic Gambling Mathematics* is my effort to distill the mathematics involved in gambling, and only that mathematics, into one place. While the final product started out as that textbook, over the course of an intensive summer spent writing (with a goal of 1000 words a day), it evolved to include more general information on the mathematics that I have found so fascinating during my years of teaching.

The text necessarily contains a large number of examples, illustrating the mathematical ideas in a range of casino games. The end of an example is indicated with a ■ symbol.

The exercises provided here are included for students, of course, and for those casual readers who would like to try their hand at some casino-related computations. Most of them either present other examples of the ideas in the chapter or ask for fairly straightforward verification of computations mentioned in the main text.

Preface to the Second Edition

Since the first edition of *BGMNBN* was published in 2014, I have had the privilege of writing and publishing three additional books on gambling mathematics [8, 9, 10]. In circling back to this book with many ideas for different material based on those books, I have been guided by the considerable thought in the gaming mathematics community about what content is desirable in an introductory course on the subject. One of the most influential recent papers on this topic is "Teaching a University Course on the Mathematics of Gambling", by Stewart N. Ethier and Fred M. Hoppe [24]. This 2020 paper reviewed a number of textbooks similar to and including *BGMNBN*, and much of the revision of this book is inspired by its comments.

Chapters 1 and 2 from the first edition have been merged into a new Chapter 1 covering both mathematical and historical background material. Part of the motivation for this change was a reviewer's excellent suggestion to get to the games earlier. Additionally, the level of mathematical complexity has been decreased in the new edition; specifically, the approach to set theory is much more informal. In writing about gambling mathematics since 2014, I have come to appreciate the advantages of a more casual approach to the underlying mathematics.

The other chapters have been updated with new examples, new exercises, and new content—one thing that is certain amidst the uncertainty of the gambling world is that there's always something new to think about and analyze mathematically. This especially applies to Chapter 4 (formerly Chapter 5) on modified casino games. There have been a number of new ideas in gambling since the first edition, and a lot of material has been replaced with fresh information. This is especially true in the area of sports gambling, which has exploded in prominence since the 2018 U.S. Supreme Court decision overturning the Professional and Amateur Sports Protection Act. With this decision, legal sports betting has spread across America, and so there is considerable new mathematics here that may be of interest to a much wider audience.

Other new material includes sections on the birthday problem (for informal gamblers) and the Monty Hall problem. The latter may not strictly be a gambling question, but it certainly stands as an excellent example of conditional probability and how it captivated the USA for a brief while in the early 1990s.

Acknowledgments

I would like to thank Callum Fraser, my editor with CRC Press, for the invitation to write a second edition of *BGMNBN*. I am especially grateful for the advice offered by the reviewers of the proposal for this edition, which have enhanced this project in ways I had not expected.

My wife Laura, though she frequently states that she doesn't understand how I do this, has been nothing but supportive as this book has taken shape, for which I am eternally grateful. It's also fun to have her as a companion as we've checked out casinos across the USA and beyond on our travels.

The Chance course that got this book started was taught for the last time in 2016, but this book has found its way into my honors course and the math majors' course Math 257: Mathematics of the Gaming Industry, which includes a field trip to Las Vegas to compare theory and practice. Some of the new material in this edition was class-tested in the 2022 section of Math 257. I would like to thank Grace Cholette, Devon Funk, Grace Hoffman, Quinlan O'Keefe, and Madeline Peterson for their suggestions that have informed and improved this revision.

The Chase the Ace image on page 85 was provided by and used with permission of Lotto Alaska. Spider craps (see page 30) was developed by Albion College student Jacob Engel during a summer research program in 2011. Funding for this project was provided by Albion's Foundation for Undergraduate Research, Scholarship, and Creative Activity (FURSCA).

Chapter 1

Fundamental Ideas

1.1 Historical Background

Gaming in antiquity

The roots of probability lie in gambling. While the mathematical foundations of probability date back fewer than 400 years, evidence of games of chance can be found among the artifacts of far-older civilizations. Dice, for example, date back many thousands of years, in much the form we know today. Dungeons & Dragons players, who use polyhedral dice with 4–20 sides, may well notice that their dice bear Arabic numerals, in contrast to cubical casino and ordinary game dice, which use dots to label sides. There is a reason for this: Dice are older than numbers [108]. Standard Arabic numerals only attained their final form around 700 CE, while cubical dice have been found dating back as far as 3000 BCE.

In China, a 1500-year-old story about looking for monkeys on mountains leads to the modern game of keno, giving keno one of the longest histories of any casino game. Legend holds that Chéung Léung (205–187 BCE), of the Han Dynasty, devised an early ancestor of keno to raise money for the army and for the defense of the capital, at a time when funds were low. This guessing game called for players to choose 8 Chinese characters from 120; the risk to the players was small and the rewards for a correct selection were great. Gamblers could buy a chance for 3 lí, and a winning ticket paid 10 taels, where 1 tael was equal to 1000 lí, for payoff odds of 10,000 for 3—about 3333 for 1. The game was a success from the start; within a few decades, the operators' wealth was said to be boundless [9].

The casting of lots to solicit the opinions of the gods as an aid to earthly decision-making eventually gave rise to the use of those devices for more recreational purposes. The Romans appear to have been the first to stage lotteries for fun; the emperor Augustus launched the forerunner of many government-run lotteries when he conducted a public works lottery with the net profit put toward funding civic improvements. The first public lottery in more modern times was held in the Netherlands in 1434, to raise money for fortifications for the Dutch town of Sluis [9].

DOI: 10.1201/9781003358183-1

Mathematical Development

The transition from simple games of chance to a mathematical theory of probability really began in the 16th century when Girolamo Cardano wrote *Liber de Ludo Aleae* or *Book on Games of Chance*. "Aleae" refers here to games played with dice [6]. The book was not published until 1663, nearly 100 years after Cardano's death—nonetheless, many of the ideas used to analyze casino games can be traced back to this work.

In this volume, Cardano gave the first mathematical treatment of *expectation*, which would come to be a fundamental idea in gambling mathematics. Looking back, we can also see that the notion of *sample space* is present, and that concept is also central to a meaningful mathematical treatment of gambling. By the end of the book, Cardano's work was showing the first signs of a theoretical, rather than experimental, approach to probability [6].

Further progress in the theory of probability can be found in Galileo's treatise on the probabilities of rolling various numbers on three standard dice. In *Sopra le Scoperte Dei Dadi*, he correctly counted the various ways to roll sums such as 9, 10, 11, and 12, and in so doing eclipsed previous incorrect reasoning that had confounded dice players [36]. Galileo's work showed, through a simple enumeration of cases, that 10 is more likely than 9 and 11 more likely than 12.

The continued progress, after Cardano and Galileo, of probability as a formal mathematical subject can be traced to a 1654 letter from Antoine Gombaud, the Chevalier de Méré, to French mathematician Blaise Pascal. In this letter, Gombaud reported his experience at two different gambling games and noted that his actual results were quite different from the results he expected based on his assessment of the probabilities. The goals of the two games he played were these:

1. To throw at least one 6 in four tosses of a fair six-sided die.

2. To throw at least one 12 (double sixes) in 24 tosses of two fair dice.

Gombaud's informal reasoning had led him to believe that his probability of winning either game was $\frac{2}{3}$, but he reportedly found that he won only slightly more often than he lost in the first game and lost slightly more often than not in the second. It is not hard to follow his reasoning: in game #1, he had 4 tries at a game with 6 possible outcomes, suggesting a $\frac{2}{3}$ chance of success; game #2 offered 24 shots at a game with 36 outcomes, leading to the same fraction.

It is also easy to find the flaw in this line of thought. Continuing game #1 for two more rolls, Gombaud would conclude that he would get at least one 6 every time he rolled a fair die six times, and it is not hard to imagine a case where this would not happen. Yahtzee and Settlers of Catan players, for example, are quite familiar with repeated inability to roll a desired number.

Pascal began a correspondence with Pierre de Fermat about these questions, which quickly grew to encompass related questions about games of chance. From these letters emerged a mathematical treatment of chance and uncertainty that laid the foundation for probability's development as a rigorous branch of mathematics. These principles would soon find application in a wide range of fields beyond gambling.

In addition to contributing to the theoretical side of probability and gambling, Pascal also helped develop the applied side of the subject. He is credited with inventing the roulette wheel while working in vain to build a perpetual motion machine.

Over the ensuing years, the foundations of probability were refined; in the early 20th century, Andrei Kolmogorov stated a set of three axioms for probability that gave probability the same logical foundation as other branches of mathematics (see page 10). Beginning with Kolmogorov's three axioms, it is possible to derive the theory of probability in complete mathematical rigor.

The Rise of Gambling in the USA

At about the same time, in 1931, Nevada paved the way for the spread of legal gambling in America by legalizing gambling within the state [108]. Over the next few decades, gambling thrived in Reno, Las Vegas, and other Nevada cities, but there was no spread of legalized casino gambling to other states until 1976, when New Jersey voters approved a measure to allow casinos in Atlantic City. Casinos opened on that city's Boardwalk in 1978 [108].

The first Native American casino in the USA was launched in a garage in Zeba, Michigan, on December 31, 1983. The Pines faced legal challenges for 18 months before being forced to close [126]. However, the push for reservation casinos continued long after the first casino shut down. In 1988, the landscape of legalized gambling in the USA was changed forever when the Indian Gaming Regulatory Act issued guidelines for regulation of casinos run by Indian tribes on reservation lands. In the years since the act was passed, Native Americans have opened casinos across America, and voter referenda in numerous other states have paved the way for state-regulated casinos. Some sort of legal gambling, including casinos, horse racing, and lotteries, is now available in 48 states, all except Hawaii and Utah.

Any attempt to write a history of poker will invariably find that much of the game's past is shrouded in mystery. Various sources trace the game's origins to the British isles, to continental Europe, and to Persia. While the history of games that we would now recognize as poker dates back to the 19th century, it exploded in popularity in the early 2000s, driven in no small part by televised Texas hold'em tournaments that used tiny cameras to show players' hole cards to the TV audience. The rise of the Internet led to interest in online poker sites, which made live gambling as close as a player's home computer. In the United States, these sites were effectively shut down on "Black Friday": April 15, 2011, when the Department of Justice seized the assets of several leading online poker sites and charged their operators with money laundering. This abruptly ended most online poker in the USA, and while the industry has rebounded somewhat, considerable legal obstacles to live online poker remain.

Another significant legal opinion that changed the face of American gambling landed in 2018, when the Supreme Court repealed the Professional and Amateur Sports Protection Act. This act had confined betting on sports to the 4 states where it was legal before the law was adopted, and its repeal left the question of sports betting in each state to that's state's lawmakers.

1.2 Mathematical Background

The mathematics behind games of chance is drawn from *probability theory* at various levels of complexity. This is appropriate, because the origins of this branch of mathematics lie with the analysis of simple games of chance. This began with a series of letters noted above, between Blaise Pascal and Pierre de Fermat in 1654, that raised and answered several questions at the foundation of probability theory while addressing questions that arose in resolving gambling games [17]. Later mathematicians have developed and expanded this topic into a rigorous field of mathematics with many important applications unrelated to gambling.

In this section, we shall outline the fundamental ideas of probability that are essential for analyzing casino games. In so doing, we will have occasion to use some of the mathematical ideas of set theory, informally. A deck of playing cards is a good illustration of the concept of a set. Considered one way, it's 52 separate things, but we may just as easily think of it as a single object.

Example 1.2.1. The set of numbers that can appear when a standard six-sided die is rolled can be written as $A = \{1, 2, 3, 4, 5, 6\}$. ∎

Example 1.2.2. A standard deck of playing cards consists of 52 cards, 13 cards in each of four suits. The 13 cards within each suit are denoted ace (A), 2 or deuce, 3, 4, 5, 6, 7, 8, 9, 10 (or T), jack (J), queen (Q), and king (K). Aces may, depending on the card game being played, be considered as either high or low. The suits are clubs (♣), diamonds (♢), hearts (♡), and spades (♠). Clubs and spades are black; hearts and diamonds are red.

Figure 1.1 shows a standard English deck. Other decks may change the design of certain cards, especially the face cards (jacks, queens, and kings) and the ace of spades.

We can think of a standard deck as a single set D containing 52 elements, which may be written systematically by listing the cards from 2 through ace in each suit and the four suits in succession alphabetically, as

$$D = \{2♣, 3♣, \ldots, A♣, 2♢, \ldots, A♢, 2♡, \ldots, A♡, 2♠, \ldots, A♠\}.$$

∎

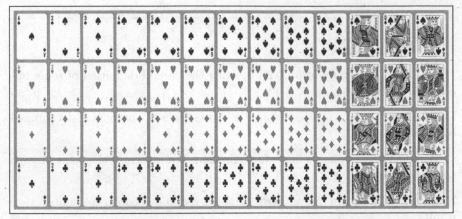

FIGURE 1.1: 52-card English deck [30].

For our work in probability, we will frequently be interested in the size of a set—that is, how many elements it has. For convenience, we introduce the following notation:

$\#(A)$ denotes the number of elements in a set A.

This is most often used when A is a finite set—while it is certainly possible to consider the size of an infinite set, such sets are uncommon in gambling mathematics and are not considered in this book.

Example 1.2.3. If A is a standard deck of playing cards, then $\#(A) = 52$. ∎

Example 1.2.4. Suppose that we roll three standard dice: one each in red, green, and blue. Let the ordered triple (r, g, b) indicate the result of the roll in the order red, green, blue—so $(2, 3, 4)$ is a different outcome from $(3, 4, 2)$. Denote by A the set of all possible ordered triples resulting from one roll. Each of r, g, and b is an integer in the range 1–6, and so we can write

$$A = \{(r, g, b) : 1 \leqslant r \leqslant 6, 1 \leqslant g \leqslant 6, 1 \leqslant b \leqslant 6\}.$$

It follows that $\#(A) = 6 \cdot 6 \cdot 6 = 216$. ∎

The challenge here is that a set of items that are of interest in gambling mathematics can be very large. If we are interested in the set A of all possible five-card poker hands, then $\#(A) = 2,598,960$, and we'd like to have a way to come up with that number without having to list all of the hands and count them. Techniques for finding the size of such large sets will be discussed in Section 2.1.

1.3 Definitions

We begin our study of probability with the careful definition of some important terms.

Definition 1.3.1. An *experiment* is a process whose outcome is determined by chance.

This may not seem like a useful definition. We illustrate the concept with several examples.

Example 1.3.1. Roll a standard six-sided die and record the number that results. ∎

Example 1.3.2. Roll two standard six-sided dice (abbreviated as 2d6) and record the sum. ∎

Example 1.3.3. Draw one card from a standard deck and record its suit. ∎

Example 1.3.4. Roll 2d6 and record the larger of the two numbers rolled (or the number rolled, if both dice show the same number). ∎

Example 1.3.5. Deal a five-card video poker hand and record the number of aces it contains. ∎

An important trait of an experiment is that it leads to a definite outcome. While we will eventually concern ourselves with individual outcomes, we begin by looking at all of the possible results of an experiment.

Definition 1.3.2. The *sample space* **S** of an experiment is the set of all possible outcomes of the experiment.

Example 1.3.6. If we consider the simple experiment of tossing a fair coin, our sample space is **S** = {Heads, Tails} or {H,T} for short. ∎

Example 1.3.7. In Example 1.3.1, the sample space is $\mathbf{S} = \{1, 2, 3, 4, 5, 6\}$. The same sample space applies to the experiment described in Example 1.3.4. ∎

Example 1.3.8. In Example 1.3.2, the sample space is

$$\mathbf{S} = \{2, 3, 4, 5, 6, 7, 8, 9, 10, 11, 12\}.$$

It is important to note that the 11 elements of **S** in this example are not equally likely, as this will play an important part in our explorations of probability. Cardano himself recognized that the proper sample space (or "circuit," to use his term) for questions involving the sum of two dice has 36, not 11, elements. ∎

Example 1.3.9. $S = \{\clubsuit, \diamondsuit, \heartsuit, \spadesuit\}$ is the sample space in Example 1.3.3. ∎

When we're only interested in some of the possible outcomes of an experiment, we are looking at subsets of S. These are called *events*.

Definition 1.3.3. An *event* A is any subset of the sample space S. An event is called *simple* if it contains only one element.

Example 1.3.10. In Example 1.3.2, the event $A = \{$The roll is an even $\#\}$ can be written $A = \{2, 4, 6, 8, 10, 12\}$. The event $B = \{$The roll is a square$\}$ is $B = \{4, 9\}$. The event $C = \{$The roll is a prime number greater than 10$\}$ is $C = \{11\}$ and is a simple event—though this is certainly not the simplest verbal description of C. ∎

Example 1.3.11. In playing video poker, suppose that your initial dealt hand is $4\heartsuit\ 5\heartsuit\ 6\heartsuit\ 7\heartsuit\ J\clubsuit$, and you discard the $J\clubsuit$ to draw a new fifth card in hopes of completing either a straight or a flush. The sample space for the draw contains the 47 remaining cards. The event corresponding to succeeding at this goal is the set A consisting of 15 cards: the 9 remaining hearts, the 3 nonheart 3s (we exclude the $3\heartsuit$ here because we already included it among the hearts), and the 3 nonheart 8s. ∎

Definition 1.3.4. Two events A and B are *disjoint* if they have no elements in common.

Example 1.3.12. In Example 1.3.3, where we draw one card from a standard 52-card deck and observe its suit, the two events $D = \{$The card is a $\diamondsuit\}$ and $H = \{$The card is a $\heartsuit\}$ are disjoint, since no card has more than one suit.

What Does it Mean to be Random?

Throughout our consideration of probability applied to gambling, we will have occasion to address the notion of *randomness*. A formal definition of "random" inherently includes humans:

Definition 1.3.5. A process is *random* if its output contains no pattern that is detectable to human observers.

This is not as satisfying a definition as we might want, but for practical purposes, it does a pretty good job of explaining what we mean when we describe something as "random." An alternate approach to randomness is this: A process is *random* if every possibility or every arrangement of its components is equally likely.

Example 1.3.13. A standard deck of playing cards is arranged in a specified order determined by the manufacturer. One such ordering begins with all of the hearts in order from ace up to king, followed by the ace through king of clubs, king through ace of diamonds, and king through ace of spades. This is about as far from random (in the sense of Definition 1.3.5) as the deck can get, and so it is important that a new deck be thoroughly shuffled before being put in play.

How thorough is "thoroughly shuffled"? Seven standard riffle shuffles, according to Dave Bayer and Persi Diaconis—provided that the shuffles are "imperfect" [63]. Perfect shuffles are those where the deck is divided exactly into two 26-card halves and the cards from each half are perfectly alternated when interlaced. These have the curious property that if the top card of the original deck remains on top throughout the process, eight perfect shuffles return the deck to its original non-random state [133].

Bayer and Diaconis showed that fewer than seven shuffles were not sufficient to randomize the deck, and that more than seven didn't improve the randomness significantly.   ■

By contrast, the street game of *Three-Card Monte* is anything but random. The setup is simple enough: a dealer flips three face-down playing cards, two red and one black, back and forth rapidly, and challenges the player to choose the black card. The machinations used by the dealer to mix the cards can be incredibly intricate.

The game could not be more simple to play: put down your money and try to pick the right card out of three. A one-in-three chance of winning is better than you get from many traditional casino games or lotteries, and if you can get just a little bit lucky, or find some kind of telltale pattern in the dealer's routine, Three-Card Monte might seem like a reasonable gamble.

It's not.

The dealer has complete control over the placement of the cards, and a skilled Three-Card Monte dealer, or *grifter*, will know where the odd card is at all times, and will control the game environment so that every outside gambler loses. Indeed, any player you see win at Three-Card Monte is a *shill*—someone who is working in league with the dealer to take all the players' money and has been tipped off to the location of the black card by the dealer's words or actions. A thorough discussion of the methods used by Three-Card Monte grifters may be found in [143].

In a modern casino, the slot machines and video poker machines are controlled by computer chips that generate thousands of numbers every second. Those numbers are determined by a proprietary algorithm that we could, in theory, exploit to predict the exact outcome of each spin of the (real or simulated) reels. Since an algorithm generates these numbers, they are properly termed *pseudorandom* numbers: not strictly random in a technical sense, but random enough for their intended use.

Example 1.3.14. The TI-58C calculator manufactured by Texas Instruments in the 1970s included a random number generator that used the following algorithm to generate a list of pseudorandom numbers [77, p. 54]:

1. Enter an initial number x_0 in the range $0 \leqslant x_0 \leqslant 199,017$.

2. Define
$$x_{n+1} = (24,298 \cdot x_n + 99,991) \mod 199,017,$$

where "mod 199,017" denotes the remainder when $24,298 \cdot x_n + 99,991$ is divided by 199,017.

3. Scale x_{n+1} as necessary to produce a random integer in a desired range. If the range is from 1 to B, the equation
$$y_{n+1} = (x_{n+1} \mod B) + 1$$

will produce an integer in that interval.

Step 2 of this algorithm will generate a list of integers in the range 0–199,016. If we begin with $x_0 = 1146$, for example, we get the sequence

$$x_1 = 83,119, x_2 = 100,937, x_3 = 180,726, x_4 = 70,234, x_5 = 74,948.$$

If we then scale these integers to the range 1–6, to represent rolls of a standard die, we would get 2, 6, 1, 5, and 3. ∎

Some attempts to cheat casinos at keno or video poker have relied on insider knowledge of these algorithms. In practice, however, doing so from the outside is so difficult that the machines are "random enough" for their intended purpose. It may well be the case that the numbers generated by the computer chip repeat with a period of 32 million, but it is equally the case that no person can hope to take advantage of this repetition. Contemporary slot machines generate numbers at a rapid rate even when they're not being played; it is only the pull of a lever or the push of a button that translates the generated number into a sequence of symbols and generates the corresponding payoff. Changing the time of the triggering event by as little as one millisecond will result in a different outcome—this is too fast for humans to exploit.

As computing power has increased, better pseudorandom number generators have been developed. The popular programming language Python uses an algorithm called the *Mersenne Twister* to generate acceptably random numbers. This algorithm is somewhat more complicated than simple modular arithmetic, but faster computers have made it possible to use more secure computational methods while still getting acceptably fast output. The twister generates numbers that run from 0 to $2^{32} - 1 = 4,294,967,295$, with a period of $2^{19.937} - 1 \approx 10^{6002}$ [101].

1.4 Axioms of Probability

In any formal mathematical system, it is necessary to specify certain statements, called *axioms* or *postulates*, which are assumed to be true without the need for proof. We may think of our axioms as the foundation on which our mathematical system is constructed, with each theorem resting atop results that are required in its proof, whether axioms or previously proved theorems. Ideally, these statements should be small in number; if there are too many axioms, proofs tend to be trivial—too many theorems are true by assumption. However, too few axioms can make early proofs difficult to construct since there is not much previous material to use in a proof. Additionally, axioms should be results that are "obvious" to reasonable observers.

Euclidean geometry was the first mathematical system founded on this axiomatic method. In Euclid's original formulation, five "common notions"—statements such as "If $a = b$ and $a = c$, then $b = c$," assumed to be true throughout all of mathematics—were joined by five postulates of a more specifically geometric nature to form the logical foundation from which all other geometric results followed. The fifth postulate ("If two lines are cut by a transversal in such a way that the interior angles on one side of the transversal add up to less than two right angles, then the lines intersect on that side of the transversal.") was considerably less intuitively obvious than the first four, and so mathematicians over the next several centuries devoted considerable effort to trying to prove it from the other axioms, without success. This was resolved in the 19th century with the development of non-Euclidean geometry, where the fifth postulate is replaced by another axiom that leads to other geometries.

The following three axioms, which were precisely formulated in 1933 by the Russian mathematician Andrei Kolmogorov, will form the foundation of our work in probability [144].

Axiom 1. *Given a sample space S, it is possible to assign to each event E a nonnegative number $P(E)$, called the **probability** of E.*

If an event is certain to occur, then its probability is 1. If an event is impossible, then its probability is 0.

Axiom 2. $P(S) = 1$.

In a given experiment, something must happen.

Axiom 3. *If $A_1, A_2, \ldots, A_n$ are pairwise disjoint events, then*

$$P(A_1 \text{ or } A_2 \text{ or } \ldots \text{ or } A_n) = P(A_1) + P(A_2) + \cdots + P(A_n).$$

In words, Axiom 3 states that the probability of any one of a set of disjoint events is equal to the sum of the probabilities of the individual events. It should be noted that the events under consideration must be disjoint—if they are not, a slightly more complicated formula called the *Second Addition Rule* may be used to find the probability that one or more of the events occurs. We shall consider this rule in Chapter 2.

As an example of how these axioms can be used to construct proofs, we shall prove a simple result called the *Complement Rule.*

Definition 1.4.1. The *complement* of an event A is the event consisting of all elements of the sample space that are not in A. We denote the complement of A by A^C.

Theorem 1.4.1. *(The Complement Rule): For any event A,*

$$P(A^C) = 1 - P(A).$$

Proof. Given any event A, we know that A and A^C are disjoint sets which, taken together, make up the entire sample space. It follows from Axiom 3 that

$$P(A \text{ or } A^C) = P(A) + P(A^C).$$

We also have

$$P(A \text{ or } A^C) = P(\mathbf{S}) = 1,$$

using Axiom 2. Combining these two equations gives

$$P(A) + P(A^C) = P(\mathbf{S}),$$

hence

$$P(A^C) = P(\mathbf{S}) - P(A) = 1 - P(A),$$

as desired. $\square$

We shall see that the Complement Rule frequently turns out to be useful in simplifying probability calculations. It can, for example, be used to resolve the Chevalier de Méré's question that launched probability theory as a branch of mathematics (see page 67).

A more significant result limits probabilities to the interval $0 \leqslant P(E) \leqslant 1$.

Theorem 1.4.2. *For any event E, $0 \leqslant P(E) \leqslant 1$.*

Proof. Axiom 1 tells us that probabilities are nonnegative, so $P(E) \geqslant 0$ and half of our conclusion is established.

We note that E and E^C, taken together, make up the sample space $\mathbf{S}$, and furthermore that E and E^C are disjoint. By Axiom 3,

$$P(E \text{ or } E^C) = P(E) + P(E^C)$$

and by Axiom 2,

$$P(E \text{ or } E^C) = P(\mathbf{S}) = 1.$$

Combining these two equations gives

$$P(E) + P(E^C) = 1.$$

If $P(E) > 1$, it follows that $P(E^C) = 1 - P(E) < 0$, an impossibility. This contradiction establishes the other half of our conclusion: $P(E) \leqslant 1$, completing the proof. $\square$

We could have chosen to use Theorem 1.4.2 in place of or in addition to Axiom 1, but since we can prove the former using the latter, and since Axiom 1 is simpler, it is the preferred choice as an axiom. It is a worthy goal, in constructing an axiomatic system, not to include anything in an axiom that can be proved from the other axioms—we say then that our axioms are *independent* of each other.

To progress further, we need to develop procedures for assigning probabilities to events. We define $P(A)$ as a ratio of the size of A to the size of the sample space $\mathbf{S}$:

Definition 1.4.2. Let $\mathbf{S}$ be a finite sample space in which all of the outcomes are equally likely, and suppose A is an event within $\mathbf{S}$. The *probability* of the event A is

$$P(A) = \frac{\text{Number of elements in } A}{\text{Number of elements in } \mathbf{S}} = \frac{\#(A)}{\#(\mathbf{S})}.$$

For convenience, we state this as a definition, although it is possible to prove this formula from Kolmogorov's axioms.

Theorem 1.4.3. *If the event B is contained in the event A, which is written $B \subset A$, then $P(B) \leqslant P(A)$.*

Proof. If $B \subset A$, then every element of B is also an element of A, but not necessarily vice versa: that is, A contains at least as many elements as B. Accordingly:

$$\#(B) \leqslant \#(A).$$

Dividing by $\#(\mathbf{S})$ gives

$$\frac{\#B}{\#(\mathbf{S})} \leqslant \frac{\#A}{\#(\mathbf{S})},$$

or

$$P(B) \leqslant P(A),$$

as desired.   □

There are several ways by which we might determine the value of $P(A)$ in practice. These methods vary in their mathematical complexity as well as in their level of precision. Each of them corresponds to a question we might ask or try to answer about a given probabilistic situation.

1. Theoretical Probability

If we are asking the question *"What's supposed to happen?"* and relying on pure mathematical reasoning rather than on accumulated data, then we are computing the *theoretical probability* of an event.

Example 1.4.1. Consider the experiment of tossing a fair coin. Since there are two possible outcomes, heads and tails, we can compute the theoretical probability of heads as $1/2$.   ∎

Example 1.4.2. If we roll 2d6, what is the probability of getting a sum of 7?

An incorrect approach to this problem is to note that the sample space is $\mathbf{S} = \{2, 3, 4, 5, 6, 7, 8, 9, 10, 11, 12\}$, and since one of those 11 outcomes is 7, the probability must be $\frac{1}{11}$. This fails to take into account the fact that some rolls occur more frequently than others—for example, while there is only one way to roll a 2, there are 6 ways to roll a 7: 1-6, 2-5, 3-4, 4-3, 5-1, and 6-1. It may be useful to think of the dice as being different colors, so that ⚁ ⚄ is a different roll from ⚂ ⚃, even though the numbers showing are the same. Table 1.1 shows all of the possible sums.

TABLE 1.1: Sample space of outcomes when rolling 2d6

+	⚀	⚁	⚂	⚃	⚄	⚅
⚀	2	3	4	5	6	7
⚁	3	4	5	6	7	8
⚂	4	5	6	7	8	9
⚃	5	6	7	8	9	10
⚄	6	7	8	9	10	11
⚅	7	8	9	10	11	12

Counting up all of the possibilities shows that there are $6 \cdot 6 = 36$ ways for two dice to land. Since six of those yield a sum of 7, the correct answer is $P(7) = \frac{6}{36} = \frac{1}{6}$. ∎

2. Experimental Probability

When our probability calculations are based on actual experimental results, the resulting value is the *experimental probability* of A. Here, we are answering the question *"What really did happen?"*

Example 1.4.3. Suppose that you toss a coin 100 times and that the result of this experiment is 48 heads and 52 tails. The experimental probability of heads in this experiment is $48/100 = .48$, and the experimental probability of tails is $52/100 = .52$. ∎

This experimental probability is different from the theoretical probability of getting heads on a single toss, which is ½. This is not unusual.

One area of gambling where experimental probability is especially prevalent is blackjack. As we will see in Chapter 5, the mathematics of blackjack can be fairly complicated, owing to the wide variety of rules that are used in various casinos and the dependence of each hand on the hands previously dealt. Many blackjack probabilities have been computed only through computer simulation of millions of hands under specified game conditions.

3. Subjective Probability

Sometimes our probability calculations are intended to answer the question *"What happened in the past?"* and use the answer to that question to predict the likelihood of a future event. A probability so calculated is called the *subjective probability* of *A*. The question we're trying to answer here is about an event that has not yet happened, instead of one that has already occurred and for which we have generated data—in this way, subjective probability can be distinguished from experimental probability.

Example 1.4.4. Perhaps the best example of subjective probability is weather forecasting. It's simply not possible to measure such parameters as temperature, humidity, barometric pressure, and wind speed and direction, plug those numbers into a (possibly) complicated formula, and have a computer generate accurate long-range weather forecasts. The inherent instability of the equations that govern weather means that small differences in the input values can lead to large differences in the predicted weather. Accordingly, there will always be some uncertainty in weather predictions. ∎

Comedian George Wallace described the imprecision of weather forecasting in the following way:

> *[The weatherman] has the only job in the world where he's never right and they say "Come back tomorrow.".*...*If you go to school, and the teacher asks you "How much is 3+4?" you can't go "Nearly 7, with a 20% chance of being 8, and maybe 9 in the low-lying areas."*

As subjective probability is, by its very nature, less mathematically precise than the other two types, we shall not consider it any further here.

The connection between theoretical and experimental probability is described in a mathematical result called the *Law of Large Numbers*, or LLN for short.

Theorem 1.4.4. (*Law of Large Numbers*) *Suppose an event has theoretical probability p. If x is the number of times that the event occurs in a sequence of n trials, then as the number of trials n increases, the experimental probability $\frac{x}{n}$ approaches p.*

Informally, the LLN states that, in the long run, things happen in an experiment the way that theory says that they do. What is meant by "in the long run" is not a fixed number of trials, but will vary depending on the experiment. For some experiments, $n = 500$ may be a large number, but for others—particularly if the probability of success or failure is small—it may take far more trials before the experimental probabilities get acceptably close to the theoretical probabilities.

Example 1.4.5. While it is possible to load a die so that the sides are not equally likely (see Example 3.1.6 for one way), it is not possible to load a coin in that way—no matter what sort of modifications are made, short of putting the same face on both sides of a coin, the probability of heads and the probability of tails will always equal ½ when the coin is tossed [38]. Nonetheless, it is customary when talking about tossing a coin to specify that the coin is "fair," and we shall do so here.

Consider the experiment of repeatedly flipping a fair coin, where the theoretical probability of heads is ½. We don't expect that every other toss will yield heads, but we do expect that the coin will land heads up approximately half the time. Table 1.2 contains data, and the corresponding experimental probabilities, from a computer simulation of a fair coin.

TABLE 1.2: Coin tossing and the Law of Large Numbers

# of tosses, n	# of Heads, x	$P(Heads)$, $\frac{x}{n}$	$P(Heads) - \frac{1}{2}$
10	3	.30	-.20
100	48	.48	-.02
1000	518	.518	.018
10,000	5044	.5044	.0044
100,000	50,039	.50039	.00039
1,000,000	499,740	.499740	-.000260
10,000,000	4,999,909	.4999909	-.0000091

While the *number* of heads tossed (column 2) is not exactly half, we observe that, as the number of trials increases, the *proportion* of tosses landing heads (column 3) gets closer to ½, and the value in the fourth column approaches 0, as the LLN suggests. ∎

1.5   Elementary Counting Arguments

Whether we are computing theoretical or experimental probabilities, we will find ourselves counting things. For simple games, it is not always necessary to invoke complicated formulas; all we need to do is be careful to make sure that every element in the set at issue is counted once and no element is counted more than once.

Example 1.5.1. A game occasionally seen in carnivals, and also present for a time on a carnival midway at the Excalibur Casino in Las Vegas, has a *barker*, or carnival employee, offering to guess your birth month. The barker is considered correct if his or her guess is within two months of your actual birthday, so if you were born in August, the barker wins on a guess of June, July, August, September, or October. You win a prize if the barker's guess is any other month. The barker typically writes down a guess before asking for your birthday, thus eliminating any opportunity for either side to profit by lying.

No matter when you were born, there are five months that, if chosen by the barker, result in a loss for you, and seven on which you win. It follows immediately that your probability of winning is $\frac{7}{12}$, and your probability of losing is $\frac{5}{12}$. ∎

While the barker can guess any month at random without affecting these probabilities, a smart strategy for the barker would look at the number of days covered by a prediction rather than the number of months, which is fixed at five. Assuming that all 365 birthdays (omitting February 29) are equally likely, the barker should choose a month to avoid catching February. A guess of any month from May through November covers three months with 31 days and two with 30, a total of 153 days. This is more days than can be covered by guessing any of the other five months, and so a barker should pick from those months exclusively. Of course, this strategy leaves the carnival open to an attack by hordes of people with February birthdays, as none of these seven months cover February in their five-month intervals.

The Big Six Wheel

The *Big Six* wheel is a large wheel (typically six feet or more in diameter; the wheel at Bob Stupak's Vegas World in the 1980s measured 30 feet across) divided into 54 sectors. Each sector bears some kind of symbol, classically a piece of currency or casino logo. Figure 1.2 shows a wheel from the Silver Legacy Casino in Reno, Nevada. These wheels are frequently placed near the perimeter of the casino floor in full view of an entrance, as if to beckon a prospective gambler with the lure of a fun and simple game.

Fun? If you're winning, probably. Every game is fun when you're winning. Simple? Yes. Players bet on one of the wheel's symbols. The wheel is spun, and if the symbol at the top of the wheel, indicated by an arrow or a leather strip called a *clapper*, matches the one the player bets on, the bet wins.

The Big Six wheel draws its name from the number of different payoffs that are possible when it's spun. A classic Big Six wheel is divided into 54 sectors, and the following symbols appear in the quantities listed in Table 1.3.

FIGURE 1.2: Big Six Wheel at the Silver Legacy Casino, Reno, NV [95].

TABLE 1.3: Big Six Wheel: Symbol frequencies

Symbol	Count
$1	24
$2	15
$5	7
$10	4
$20	2
Logo A	1
Logo B	1

In Figure 1.2, Logo A is the "Joker" space and Logo B is labeled "Silver Legacy". At the Venetian Casino in Las Vegas, the two logos on the wheel were a joker and an American flag. Down the Strip at the Cosmopolitan, the logo slots were labeled "Art" and "Music."

Calculating probabilities on the Big Six wheel is a simple matter of counting the sectors containing the symbol on which you have bet and dividing by 54. Accordingly, we have the probabilities listed in Table 1.4.

TABLE 1.4: Big Six Wheel probabilities

Symbol	Probability
$1	$24/54$
$2	$15/54$
$5	$7/54$
$10	$4/54$
$20	$2/54$
Logo A	$1/54$
Logo B	$1/54$

Roulette

> *My lucky number is 4 billion. That doesn't come in real handy when you're gambling.*
>
> —Mitch Hedberg, *Strategic Grill Locations*

In the game of *roulette*, a small ball is spun on a rotating wheel and comes to rest in a numbered pocket on the wheel. *European* roulette uses a wheel divided into 37 pockets, numbered 1 to 36 and 0. The 0 is colored green, and the remaining 36 slots are split evenly between red and black. In *American* roulette, a 38th pocket, colored green and numbered 00, is added—see Figure 1.3.

FIGURE 1.3: American roulette wheel [29].

Players select numbers from the wheel and are paid if the ball lands in a pocket bearing one of their numbers. A variety of bets, at a range of payoffs, is available to players—see Figure 1.4 for an illustration of the betting layout.

Table 1.5 lists the standard bets and their corresponding payoffs. Wagers made on the grid of numbers, covering 6 or fewer numbers, are called *inside*

FIGURE 1.4: American roulette layout [28].

bets, while bets on 12 or more numbers, placed at the edges of the layout, are known as *outside* bets.

TABLE 1.5: Roulette bets and payoffs

Bet	# of numbers	Payoff
Straight	1	35 to 1
Split	2	17 to 1
Street	3	11 to 1
Corner	4	8 to 1
Basket	5	6 to 1
Double street	6	5 to 1
Dozen	12	2 to 1
Even-money	18	1 to 1
Double column	24	½ to 1

- *Straight* bets cover a single number.

- *Split* bets are made on two adjacent numbers on the layout by placing a chip on their common line segment. Except for 0 and 2, numbers that are adjacent on the layout are not adjacent on the wheel.

- A *street* bet is made on a row of three consecutive numbers, for example 28, 29, and 30, by laying that chip at the edge of the row.

- *Corner* bets may be made on any block of 4 numbers forming a 2×2 square, by placing the chip at the center of that block.

- The *basket* bet is available only on American roulette wheels and can only be made on the five-number combination 0, 00, 1, 2, and 3. A player makes this bet by putting chips down at the edge of that block of five numbers, either at the corner where 0 and 1 meet or the corner where 00 meets 3.

- *Double street* bets cover 6 numbers, 2 adjacent rows, and are made by placing chips on the edge of the grid of numbers where two numbers meet, as where 24 and 27 come together. This double street bet would cover the numbers 22 through 27.

- A *dozen* bet may be made on the numbers from 1 to 12, 13 to 24, or 25 to 36, or on any of the three columns on the betting layout.

- An *even-money* bet may be made on odd, even, red, black, low (1–18), or high (19–36) numbers. Every even-money bet covers exactly 18 of the 38 numbers; 0 and 00 are neither even nor low.

- *Double column* bets, which are not offered at many casinos, cover two adjacent columns of 12 numbers each.

As a convenient shortcut, it should be noted that, except for the basket bet, the payoff for a bet on n numbers is $\dfrac{36 - n}{n}$ to 1. A winning player's initial bet is returned with the payoff.

The probability of winning a roulette bet is easy to compute:

$$P(\text{Win}) = \frac{\text{\# of numbers covered by the bet}}{\text{\# of numbers on the wheel}}.$$

In European roulette, the denominator is 37; in American roulette, it is 38. This formula produces the probabilities recorded in Table 1.6.

We can easily see that European roulette gives slightly higher probabilities of winning the same bet, so if you have a choice between European and American roulette, pick European. It's better for the player, though as we shall see, both games still favor the casino. Casinos which offer both American and European roulette tables frequently extract a price from gamblers in the form of higher bet minimums at European tables, which require players to risk more money if they want a game with a better chance of winning.

In 2016, the Venetian Casino in Las Vegas introduced *Sands Roulette*. This game added a 39th pocket to an American roulette wheel, which was colored green and bore an "S" for Sands, the parent company of the Venetian. The S space acts like a third zero for betting and payoff purposes. It is represented on the layout by a rectangle above the 0 and 00 spaces, so it also belongs to no column. Basket bets are available on this layout.

The following year, 000 spaces appeared on some roulette wheels along the Strip, at New York–New York and Planet Hollywood, and the Luxor added a

TABLE 1.6: Roulette probabilities

Bet	P(Win) European	American
Straight	$1/37$	$1/38$
Split	$2/37$	$2/38$
Street	$3/37$	$3/38$
Corner	$4/37$	$4/38$
Basket	—	$5/38$
Double street	$6/37$	$6/38$
Dozen	$12/37$	$12/38$
Even-money	$18/37$	$18/38$
Double column	$24/37$	$24/38$

39th pocket bearing its pyramid logo. Triple-zero roulette wheels have since been installed at numerous Strip casinos. The payoffs remain the same: $\dfrac{36 - n}{n}$ to 1 except for the basket bet, but the probability of winning drops [10].

California state law forbids the use of a roulette wheel or dice as the sole determiner of the outcome of a game of chance, requiring that the outcomes of all casino games be determined by cards. Nonetheless, casinos in San Diego County have found a number of innovative ways to use a roulette wheel or other gaming devices to run a game that is mathematically equivalent. At the Barona Casino, a single-zero wheel is spun and 3 cards from a 37-card deck, numbered 0 to 36, are dealt to the table. If the number spun is in the range 0–12, the first card is flipped over and used as the result of the game. Similarly, the second card is used if the wheel comes up 13–24 and the third if the number on the wheel is 25–36.

Craps

Craps is a popular casino game played with two six-sided dice. A game begins when the shooter rolls the two dice, which is called the *come-out* roll. If the come-out roll is 7 or 11, this is an immediate win for the shooter; if 2, 3, or 12, an immediate loss.

If the come-out roll is any other number, that number becomes the *point*. The shooter then continues rolling until he either rolls the point again, or rolls a 7. All other rolls are disregarded for the purposes of resolving this main bet; there are several other betting options that can be chosen on individual rolls. The shooter wins if he re-rolls the point before a 7, and loses if he rolls a 7 first.

There is a large collection of wagers available to a craps bettor. Figure 1.5 depicts all of the bets available on a standard craps layout. A full-size craps table would include a second betting field for come, pass, and field bets, placed symmetrically to the right of the illustration to duplicate the betting options at the other end of the table.

FIGURE 1.5: Craps layout [60].

Example 1.5.2. We shall focus first on the *Pass* and *Don't Pass* bets. A Pass bet is a bet that the shooter will win, and a Don't Pass bet is a bet that the shooter will lose. For the Pass bet, there are six ways to roll a 7 and two ways to roll an 11 on the come-out roll, so $P(\text{Win on the come-out roll}) = \dfrac{8}{36}$. If a point is established, the probability of an eventual win depends on the value of the point. Once the point is known, the only rolls that matter are those resulting in that number or in a 7. For example, if the point is 9, all we look at are the four ways to roll a 9 and the six ways to roll a 7, so there are ten rolls that will resolve the pass/don't pass question. The probabilities of winning with each point are shown in Table 1.7.

TABLE 1.7: Craps: Win probabilities once points are set

Point	Ways to roll	$P(\textbf{Win})$
4 or 10	3	$3/9$
5 or 9	4	$4/10$
6 or 8	5	$5/11$

The probability of each number becoming the point is equal to the number of ways to roll that point divided by 36, so, for example, the probability of first establishing the number 4 as a point and then successfully making that point is $\dfrac{3}{36} \cdot \dfrac{3}{9} = \dfrac{1}{36}$. For the other points, we have the figures in Table 1.8.

TABLE 1.8: Craps win probabilities

Point	P(Point rolled)	P(Point wins)	P(Win on this point)
4 or 10	$3/36$	$3/9$	$1/36$
5 or 9	$4/36$	$4/10$	$2/45$
6 or 8	$5/36$	$5/11$	$25/396$

By adding up the seven probabilities so far determined, we find that the probability of winning a pass line bet is approximately

$$p = \frac{8}{36} + 2 \cdot \frac{1}{36} + 2 \cdot \frac{2}{45} + 2 \cdot \frac{25}{396} = \frac{244}{495} \approx .4929.$$

∎

If there were no other rules, it would follow that the probability of winning a Don't Pass bet would be $q = 1 - p \approx .5071$, which is greater than 50%—and this is not in the casino's interests, for the gambler would have an advantage.

To prevent this, most casinos have adopted a "Bar 12" rule for the come-out roll, which means that if the shooter rolls a 12 on the come-out roll, Pass bets lose but Don't Pass bets push (tie) instead of winning. With this $\frac{1}{36}$ removed, the probability of winning on a Don't Pass bet is $p = .4792$. Some casinos bar the 2 rather than the 12; since 2 and 12 are equally likely rolls, the probabilities are the same.

A standard craps layout also includes spaces labeled "Come" and "Don't Come." These wagers act and are paid exactly like Pass and Don't Pass, but may be made by a bettor at any time, without waiting for the previous point to be resolved. When a Come or Don't Come bet is made, the next roll of the dice functions as a come-out roll. If the bet is not resolved immediately by the roll of a 2, 3, 7, 11, or 12, then the chips are moved to a numbered space on the layout just above the "Come" space bearing the point rolled, where the bet can be monitored by casino personnel until it is resolved.

If Pass/Come and Don't Pass/Don't Come were the only bets available at craps, there would be a lot of downtime without much wagering, as it can potentially take many rolls before the pass/don't pass question is resolved. To maintain the level of excitement and encourage additional wagering, a craps layout has many more betting options available to players.

One class of craps wagers is *one-roll* bets, which are simply bets on what the next roll of the dice will be and thus are quickly resolved.

Example 1.5.3. A very simple one-roll bet is the *Eleven* bet, which wins if the next roll is an 11 and loses otherwise. There are two ways to roll an 11—6-5 and 5-6—so the probability of winning the Eleven bet is $\frac{2}{36}$, or $\frac{1}{18}$. ∎

In practice, a roll of 11 is frequently announced with a shout of "Yo!"—this is short for "yo-leven," which is used to distinguish "eleven" from the similar-sounding "seven." A longtime craps superstition holds that saying "seven" at the table is bad luck, so avoiding anything that sounds like that number eliminates confusion.

It should be noted that the craps layout depicted in Figure 1.5 gives the payoff for the Eleven bet as "15 for 1," which means that a winning bettor's wager is included as part of the payoff, so that the effective payoff is only 14 to 1. "15 for 1" sounds like a bigger prize than the identical payoff of "14 to 1," which may explain this word choice.

Example 1.5.4. The multiple "C" and "E" spots located to the left of the center betting section in Figure 1.5 stand for Crap and Eleven, and are placed on the felt in such a way that they point at players who gather at the table. By using these betting spots, dice dealers can more easily keep track of which bets belong to which bettor. C and E are used for making the *Craps & Eleven* one-roll bet, which combines the "Any Craps" (2, 3, or 12) bet with the Eleven bet. Any Craps pays off at 8 for 1 if the next roll is 2, 3, or 12; these are the numbers that lose for the Pass better on the come-out. The Eleven bet pays off at 15 for 1 if the next roll is an 11.

In practice, a player's C&E wager is split evenly between the C and E spaces; if this is not possible, as for example with a $5 wager, the chip or chips are placed between the circles, and any fractional payoff is rounded down, in the casino's favor.

The probability of winning the C&E bet can be calculated by adding up the number of ways to roll a 2, 3, 11, or 12, and is simply

$$p = \frac{1+2+2+1}{36} = \frac{6}{36} \approx .1667.$$

■

Blackjack

Blackjack is one of the most popular casino games, in part because the gambler plays an active role in the game by his or her choices of how to play the cards that are dealt. The goal in blackjack is to get a total closer to 21 than the dealer, without going over 21. Aces count either 1 or 11, at the player's discretion. Face cards count 10, and all other cards count their rank. A full examination of blackjack may be found in Chapter 5; we examine here some simple questions that can be answered with the mathematics we have developed so far.

Example 1.5.5. Blackjack hands start with two cards dealt to each player. In a blackjack game dealt from a double deck—two standard decks shuffled together—find the probability that your first card is an ace.

There are 8 aces in a double deck, and 104 cards in all, so

$$P(\text{Ace}) = \frac{8}{104} = \frac{1}{13}.$$

∎

Example 1.5.6. In a double-deck game, suppose that you have been dealt two 10s, for a total of 20. The dealer's first card is an ace. What is the probability that his second card will be a 10-count card (10, jack, queen, or king) that brings his total to 21 and beats you?

There are 101 cards as yet unaccounted for. Each deck starts out with 16 ten-count cards, for a total of 32 at the hand's start, and two of them are in your hand and so cannot be drawn by the dealer. This leaves 30 in the deck, and we have

$$P(\text{Dealer 21}) = \frac{30}{101} \approx .2970,$$

or slightly less than 30%.

∎

Poker

Whether played around dining room tables, in college dorm rooms, or at casino poker rooms like the one at the aptly-named Poker Palace in North Las Vegas, Nevada (Figure 1.6), the goal in many poker games is to make the best 5-card hand, however "best" may be determined by the rules of the game.

FIGURE 1.6: The Poker Palace casino in North Las Vegas, Nevada [138].

5-Card Poker Hands

The following 5-card hands, listed here from highest to lowest ranked, are commonly recognized.

1. The highest-ranked 5-card poker hand is a *royal flush*, which consists of the ace, king, queen, jack, and 10 of the same suit. If two players hold royal flushes in the same hand, they split the pot—suits have no standing in most poker games.

2. Royal flushes may also be thought of as the highest-ranking *straight flush*. A straight flush includes 5 cards of consecutive ranks in the same suit. Aces can count as high, which results in a royal flush, or low, as in 5432A♡, the lowest-ranked straight flush. In a showdown between two straight flushes, the hand with the highest card wins. If two players have the same cards in different suits, they split the pot.

   The straight flush, or "sequence flush", to use its original name, is the newest widely recognized 5-card poker hand, arriving on the scene in the late 19th century. The hand is mentioned in an 1875 book on poker [147], but an 1876 edition of *Hoyle's Book of Games* fails to mention either straight flushes or straights among its ranking of poker hands [32].

3. *Four of a kind* is the third highest hand. These hands consist of all 4 cards of one rank together with an unmatched card.

   Prior to the recognition of the straight flush, including the royal flush, as a separate hand, a player holding 4 aces, or 4 kings and an ace, knew that his hand was unbeatable and could bet accordingly. This was thought by some gamblers and poker writers to be contrary to the spirit of gambling [62, 91]. Even if a player holds a royal flush, there can be a very small chance, depending on the game, that one or more other players may also hold one, so the modern ranking of poker hands has no unbeatable hand.

4. A *full house* includes 3 cards of one rank and 2 cards of another rank.

   If two players hold full houses, the hand with the higher-ranked three-of-a-kind wins the showdown, so 99922 beats 444*AA*. The pair makes the hand a full house, of course, but otherwise, it adds no value [62]. Since the pair is immaterial in the rankings, full houses are sometimes described briefly by the rank of the three-of-a-kind, such as "aces full" for any hand of the form *AAAxx*.

   In [92], Richard A. Proctor correctly notes that 3 aces and a pair of 3s beats 3 aces and a pair of 2s, but if both were to occur in the same hand of 5-card draw poker, there are larger questions to be answered than who holds the better cards. In Texas hold'em poker, such a showdown is possible.

5. A *flush* is a hand in which all 5 cards are of the same suit, but are not in sequence. When two or more players hold a flush, the hand with the highest card is the winner.

6. *Straights* have 5 cards in sequence, but not all in the same suit. As with straight flushes, an ace can be high or low, but straights are not allowed to go "around the corner", counting the ace as both high and low, as in $KA234$. A "Broadway straight" is the highest possible straight: $AKQJT$.

As noted above under straight flushes, straights were not always accepted as a poker hand. An 1866 work by William B. Dick made mention of straights, or "rotations", but while asserting that "straights are not considered in the game", allowed that "they are played in some localities, and it should always be determined whether they are to be admitted at the commencement of the game" [18]. The book did, however, go on to support the recognition of the straight flush as the highest-ranking hand, citing the inadvisability of an unbeatable hand.

In a poker game played with 5-card hands dealt from a standard 52-card deck, any straight must contain either a 5 or a 10. This fact has some implications for stud poker games where each player can see some of their opponents' cards.

7. *Three-of-a-kind* hands are made up of 3 cards of one rank and two odd cards which match neither the three-of-a-kind nor each other.

8. *Two pairs* ranks next: a hand with two pairs of cards of different ranks and a fifth card not matching either pair.

Two-pair hands are described by the rank of the higher pair, such as "9s up", for a pair of 9s and any lower pair. In a showdown between 2 two-pair hands, the highest pair wins. If both hands have the same higher pair, then the hand with the higher lower pair wins, for example, $KK88T$ beats $KK44Q$. Should the hands have the same pairs, the odd card determines the winner, so $7733A$ beats 77334.

9. A *pair* includes 2 cards of the same rank and 3 unmatched cards. Ties between pairs are broken by looking at the highest unpaired cards in each hand.

10. A *high card* hand consists of 5 unmatched cards, not in sequence, covering at least 2 suits. This is the lowest-ranked hand.

These relative hand rankings are used in any form of poker using a standard deck. If wild cards are added to the deck, as when using a 53-card deck including a joker, some slight variations in hand frequency are introduced (see page 190).

There are numerous variations on poker's fundamental idea, which is to achieve as high-ranking a hand as possible within the rules of the game at play. In a variation called *lowball*, the goal is to achieve the lowest possible hand with straights and flushes often excluded from consideration, so "high-ranking" in a lowball game would mean "weakest" according to this list.

Texas hold'em

Texas hold'em is a form of poker whose popularity skyrocketed in the early 2000s, in large part due to its presence on television. The game is played as follows:

- Two cards—the *hole* or *pocket* cards—are dealt face down to each player. This is followed by a round of betting.

- Three cards—the *flop*—are dealt face up to the center of the table. These three cards and the two that follow are *community* cards, and are used by all players, together with their two hole cards, to make the best possible five-card poker hand. A second round of betting follows the flop.

- A fourth community card—the *turn*—is dealt face up to the table, and a third round of betting follows.

- The final community card—the *river*—is dealt to the table, completing the hands. A final round of betting takes place, and at the end, the player with the highest five-card poker hand, made from the community cards and their two hole cards, remaining in the game wins the pot.

The fall of the cards at Texas hold'em suggests many probability questions, some of which can be answered using no more than simple counting. Televised poker, whose popularity may be traced in no small part to the development of tiny cameras that allow home viewers to see the players' hole cards, usually includes the probability of each player winning the hand, which is calculated by a brute-force computer simulation. All possible choices of the cards remaining to be dealt are considered, and the winning hand for each arrangement is determined. Simple counting of the number of times each player wins gives the experimental probabilities shown on the screen.

Example 1.5.7. Suppose that your hole cards are the 4♣ and 5♠, the flop is 6♦ 2♡ 5♦, and the turn is the 6♡. What is the probability that the river will complete a full house?

To pull a full house, the river must be one of the two remaining 5s or the two 6s. Since there are 4 such cards, while 46 cards remain in the deck, you have probability $\frac{4}{46} \approx .0870$ of drawing into a full house. ■

We have carried out this calculation without concern for what cards other players may hold and thus are unavailable to be drawn on the river. If all of the 5s and 6s have been dealt out, then your probability of completing your full house is 0, but if none of them is in other players' hands, then your chances have increased past .0870. In light of the information we have, .0870 is the best estimate for $P(\text{Full house})$, and if we were to compute the probability for every possible game situation, we would find that the *average* probability of a full house would be .0870.

Example 1.5.8. In the example above, what is the probability that you will complete a straight on the river?

To fill in your straight, you need to draw a 3. Since there are four 3s as yet unaccounted for, the probability of getting a straight is also $\frac{4}{46}$. ∎

There is considerable literature on Texas hold'em that takes the game's mathematics into account. One of the best books is *Introduction to Probability with Texas Hold'Em Examples* by Frederic Paik Schoenberg [106]. A good portion of the game's strategy is when to make a bet or fold a given hand, and most experts would agree that the hand above should be folded before the flop, as two low cards, even consecutive ones, of different suits comprise a very weak starting hand.

Example 1.5.9. Find the probability that your hole cards are the same suit—"suited," in the game's lingo.

The first card doesn't matter; what we need to compute is the probability that your second card matches its suit. Again, with no knowledge about the other players' holdings, there are 12 cards of your suit remaining out of 51 overall, so the probability of suited hole cards is $\frac{12}{51} \approx .2105$. ∎

1.6 Exercises

Answers to starred exercises begin on page 285.

1.1.* A *complete bet* at roulette is a collection of bets that cover every possible way, through split, street, corner, and double street bets, to make an inside bet on a single number. The cost of a complete bet depends on the number selected, which affects the number of some of the compound bets that are available. For example, a complete bet on the number 23 includes a single-number bet on 23; four split bets covering 23 with one other adjacent number; one street bet on 22, 23, and 24; four corner bets; and two double street bets. At $1 per bet, this wager would cost $12 altogether.

a. Assuming an American roulette wheel, find the probability of hitting a winning number on a complete bet on 23.

b. Referring to Figure 1.4, how many separate $1 bets must be made to make a complete bet on the number 33? Find the probability of hitting a winning number, on an American roulette wheel, for this bet.

1.2.* Some casinos pay a small bonus on certain blackjack hands—for example, a hand containing three 7s, totaling 21. In a double-deck blackjack game, find the probability of being dealt three 7s.

1.3.* Find the probability of being dealt a pair as your hole cards in Texas hold'em.

1.4.* The following lyrics are from the song "Transistor Radio" by British comedian Benny Hill:

> *Last night I held her little hand,*
> *It made my poor heart sing.*
> *It was the sweetest hand I'd held,*
> *Four aces and a king.*

Explain why the probability of being dealt four aces and a king is the same as that of being dealt a royal flush.

1.5.* Considering only jacks, queens, and kings as face cards, find the highest and lowest possible ranks for a poker hand consisting of 5 face cards.

1.6.* *Mini-craps* is a version of craps that is played on a smaller table and frequently offers simpler betting options. One version of mini-craps removes the four line bets: pass, don't pass, come, and don't come. These are replaced with the simpler "Over 7" and "Under 7" bets, where the gambler simply wagers on the value of the next roll of the dice. These wagers pay off at even money. Find the probability of winning the Over 7 bet, which wins if the subsequent roll is 8 or higher.

Spider craps

Spider craps is a game variation that uses eight-sided dice rather than six-sided dice. It was originally proposed by Richard Epstein in 1977 and called "sparc" [22]. The game was later developed more fully by Jacob Engel. In order to design a game that will be attractive both to players and to casinos, certain modifications are made that are nonetheless consistent with the rules of the original game. The following exercises explain and explore some of the rules of spider craps.

1.7.* On the come-out roll in spider craps, a Pass bet wins instantly if the shooter rolls a 9 or a 15 and loses if the roll is a 2, 3, or 16. Find the probability of winning a Pass bet and the probability of losing a Pass bet on the come-out roll.

1.8. In spider craps, once a point (4–8 or 10–14) has been established on the come-out roll, a Pass line bet wins if the shooter re-rolls that number before rolling a 9. By mimicking the calculations leading to the table on page 23, compute the probability of winning a Pass line bet at spider craps.

1.9.* If no rolls are barred on the come-out roll, then a Don't Pass bet at spider craps will have a probability of winning that exceeds .5. Show that barring only the 16 (analogous to standard craps), or the 2 and 16 together, still results in a bet with an advantage for the player.

1.10. Show that if both 2 and 3 are barred on the come-out roll, then the spider craps Don't Pass bet has a probability of winning that is under 50%. This is the rule in force in spider craps.

Chapter 2

Combinatorics and Probability

Combinatorics is the branch of mathematics that studies counting techniques. In many applications of Definition 1.4.2, $P(A) = \dfrac{\#(A)}{\#(\mathbf{S})}$, the sheer number of elements comprising an event or the sample space is far too large to be counted one by one. When computing probabilities, we seldom have need to consider each of these simple events individually; we are usually only interested in how many there are. This chapter will identify some common counting challenges that arise in gambling mathematics and suggest several techniques to compute the numbers required in an efficient way.

2.1 Advanced Counting Arguments

Fundamental Counting Principle

Frequently in gambling mathematics, we find ourselves considering the number of ways in which several events can happen in sequence. If we know the number of ways that each individual event can happen, elementary combinatorics tells us that simple multiplication can be used to find the answer.

Theorem 2.1.1. *(The Fundamental Counting Principle) If there are n independent tasks to be performed, such that task T_1 can be performed in m_1 ways, task T_2 can be performed in m_2 ways, and so on, then the number of ways in which all n tasks can be performed successively is*

$$N = m_1 \cdot m_2 \cdot \ldots \cdot m_n.$$

That the FCP is a reasonable result can be easily seen by testing out some examples with small numbers and listing all possibilities—for example, when rolling 2d6, one red and one green, each die is independent of the other and can land in any of six ways. By the FCP, there are $6 \cdot 6 = 36$ ways for the two dice to fall, and this may be confirmed by writing out all of the possibilities.

As an example of how the Fundamental Counting Principle is used, we can show that the number of possible events associated with a probability experiment is exponentially greater than the number of elements. Specifically, we have the following result.

DOI: 10.1201/9781003358183-2

Theorem 2.1.2. *If $\#(S) = n$, then there are 2^n events that may be chosen from S.*

Another way to state Theorem 2.1.2 is to say that a set with n elements has 2^n subsets. For example, if we toss a coin, we have $S = \{H, T\}$. Since we have $\#(S) = 2$, we expect $2^2 = 4$ events. The complete list of events contained within S is $\varnothing, \{H\}, \{T\}, \{H, T\}$, where $\varnothing$ denotes the *empty set*: an event containing no elements. This list indeed numbers 4.

Proof. Let A be an event within the sample space S. For every element x in S, either x is part of A or x is not part of A. That is, there are 2 possibilities for each element: "Yes, it is in A" or "No, it is not in A." Every arrangement of the n possible yes and no answers corresponds to a different subset of S, and thus to a different event. For example, choosing "no" for every element gives the empty event, and at the other extreme, choosing "yes" each time gives S itself.

There are n choices to be made, and two options for each of the n choices. Therefore, by the Fundamental Counting Principle, there are

$$\underbrace{2 \cdot 2 \cdot \ldots \cdot 2}_{n \text{ terms}} = 2^n$$

possible events. $\qquad\square$

Example 2.1.1. If we consider the experiment of drawing a card from a standard deck, then $\#(S) = 52$, and there are $2^{52} = 4,503,599,627,370,496$ different events we could consider. Most of these are of limited mathematical interest—for example, the event "Draw any card except the 3♣ or 7♡," with 50 elements, comes up only very rarely.

On the other hand, an event such as "Draw any face card or 10," which has 16 elements, has considerable importance when you've been dealt an ace as your first card in a hand of blackjack. ∎

Old-style mechanical three-reel slot machines often had 20 symbols per reel. When the handle was pulled, each of the reels would spin independently before stopping with one symbol on the central payline. By the Fundamental Counting Principle, there are $N = 20 \cdot 20 \cdot 20 = 8000$ different arrangements of three symbols.

Modern computerized slot machines are considerably more intricate. On a three-reel or five-reel electromechanical slot machine, the position of each reel after the handle is pulled (or the SPIN REELS button is pressed) is determined by a randomly generated number, and the number of symbols on the reels need not bear any direct relationship to the total number of possible random numbers. Some reel configurations, such as a blank on all three reels, may correspond to hundreds of random numbers.

On a video slot machine, the screen displays simulated reels, and the computer can process winning and losing combinations on dozens, even hundreds, of paylines that criss-cross the screen connecting symbols. As with computerized reel slots, a generated random number determines the arrangement of symbols on the virtual reels.

Example 2.1.2. The *Carnival of Mystery* video slot machine, manufactured by International Game Technology, uses a five-reel screen with three symbols displayed on each reel. The game can be played with any symbol on each reel combining with any symbol on each other reel to form a payline. A payline is built by choosing one displayed symbol from the three on each reel, and so the total number of paylines is $3^5 = 243$. On a penny machine, all 243 lines can be played for one wager of 25¢. ∎

A special case of the Fundamental Counting Principle arises when we consider the number of ways to arrange a set of n elements, with no repetition allowed, in different orders. The first element may be chosen in n ways, the second in $n - 1$, and so on, down to the last item, which may be chosen in only one way. The total number of orders for a set of n elements is thus $n \cdot (n-1) \cdot (n-2) \cdot \ldots \cdot 3 \cdot 2 \cdot 1$. This number is given a special name, n *factorial*.

Definition 2.1.1. If n is a positive integer, the *factorial* of n, denoted $n!$, is the product of all of the positive integers up to and including n:

$$n! = 1 \cdot 2 \cdot 3 \cdot \ldots \cdot (n-1) \cdot n.$$

$0! = 1$, by definition.

It is an immediate consequence of the definition that $n! = n \cdot (n-1)!$. Factorials get very big very fast. $4! = 24$, but then $5! = 120$ and $6! = 720$. 10! is greater than 1 million, and 52!, which is the number of different ways to arrange a standard deck of cards, is approximately 8.066×10^{67}. In the 1970s and 1980s, the largest factorial that could be computed on a standard scientific calculator was $69! \approx 1.711 \times 10^{98}$, a limit well-known to many people whose interest in mathematics was generated in part by playing with early calculators.

Factorials may be computed on the Texas Instruments TI-84+ calculator by following these steps:

- Enter n.

- Press $\boxed{\textbf{MATH}}$ to bring up the Math menu.

- Scroll to the right to **PRB**. This brings up the Probability submenu.

- Select option **4: !** and press $\boxed{\textbf{ENTER}}$.

- Press $\boxed{\textbf{ENTER}}$ again.

Where Order Matters: Permutations

Definition 2.1.2. A *permutation* of r items from a set of n items is a selection of r items chosen so that the order matters.

For example, ABC is a different choice of three alphabet letters than CBA. It should be noted that "order" may appear in several forms. One way to determine whether or not order matters in making a selection is to ask if different elements of the selection are being treated differently once they are chosen.

Example 2.1.3. Suppose that a baseball team includes 15 players, each of whom can play any of the 9 positions. If we select nine players to start a game, the assignment of players to positions is a different treatment of the individual players. It would be correct to conclude that the order of selection matters—the same subset of nine players can be assigned to the nine positions in many ways.   ∎

We are usually not as interested in a list of all of the permutations of a set as in how many permutations there are. The following theorem allows easy calculation of that number, which is denoted $_nP_r$.

Theorem 2.1.3. *The number of permutations of r items chosen from a set of n items is*

$$_nP_r = \frac{n!}{(n-r)!}.$$

Proof. There are n ways to select the first item. Once an item is chosen, it cannot be chosen again, so the second item may be chosen in $n-1$ ways. There are then $n-2$ items remaining for the third choice, and so on. By the Fundamental Counting Principle, we have

$$_nP_r = n \cdot (n-1) \cdot \ldots \cdot (n-r+1).$$

Multiplying the right-hand expression by $1 = (n-r)!/(n-r)!$ gives

$$_nP_r = n \cdot (n-1) \cdot \ldots \cdot (n-r+1) \cdot \frac{(n-r)!}{(n-r)!}$$
$$= \frac{n \cdot \ldots \cdot (n-r+1) \cdot (n-r) \cdot \ldots \cdot 3 \cdot 2 \cdot 1}{(n-r)!}$$
$$= \frac{n!}{(n-r)!}.$$

□

The number $_nP_r$ can be calculated easily on a TI-84+ calculator as follows:

• Enter n.

- Press $\boxed{\text{MATH}}$ to bring up the Math menu.

- Scroll to the right to **PRB**. This brings up the Probability submenu.

- Select option **2: nPr** and press $\boxed{\text{ENTER}}$.

- Enter r and press $\boxed{\text{ENTER}}$.

Horse racing

There are very few casino games where the order of events matters and permutations must be counted. One place where order is important when gambling is in horse racing. The *trifecta* wager requires a gambler to pick the first three horses to finish a race, in order. It's not enough simply to pick the right three horses, it's also necessary to identify which horse wins, which horse comes in second (or *places*), and which is third (the *show* horse). This, of course, requires considerable skill—or luck.

Example 2.1.4. In recent years, the Kentucky Derby, the first race in thoroughbred racing's Triple Crown, has been restricted to 20 horses, determined originally by lifetime winnings and beginning in 2013 by a point system taking the results of preliminary races into account. If you pick your trifecta ticket at random, what is the probability of choosing the winning three horses correctly?

There are

$$_{20}P_3 = 20 \cdot 19 \cdot 18 = 6840$$

different ways to choose the top three horses, so your chance of winning with a single ticket is $\dfrac{1}{6840}$. ∎

Note that the calculation of this probability assumes that all possible top-three outcomes are equally likely, which may not be appropriate to every race. If one horse in the field is very heavily favored (see Example 2.2.3 for one such race), then it may be wise to slot that horse into the winner's spot and consider only the place and show positions for your wager. It may be more sensible, from a purely mathematical perspective, to consider any choice of horses for those two slots as equally likely.

If the trifecta wager seems too risky for your tastes, the *exacta* bet, which requires only that you pick the top two horses in order, may be more suitable. In the example above, with 20 horses starting, the chance of picking the top two finishers in order correctly is

$$\frac{1}{_{20}P_2} = \frac{1}{380}.$$

Going in the other direction, some racetracks offer a *superfecta* wager, which pays off if you correctly pick the top *four* finishers. There are

$$_{20}P_4 = 20 \cdot 19 \cdot 18 \cdot 17 = 116,280$$

different ways to arrange 4 horses out of 20 in order, so your chance of winning on a single superfecta ticket is $\frac{1}{116,280}$. It follows that winning superfecta tickets are rather rare and thus are likely to pay off handsomely if the field is relatively large.

How handsomely do winning superfecta tickets pay off? Racetracks use a method called *pari-mutuel* wagering, where the money bet on a given proposition is pooled. As more and more races are run without a winning superfecta wager, the pool accumulates money. The racetrack retains a portion of the wagers, which can run as high as 20%, and the remainder is equally divided among everyone holding a winning ticket. In managing bets this way, the track assures that its profit is taken off the top before any payoffs are made. Similar separate pools are established for other bets, such as trifecta and exacta bets. The accumulated pool is divided among the winning bettors, so the odds you take when you bet may not be the odds at which you are paid; a winning ticket is paid in accordance with the odds in force at the time that betting for the race ends.

Example 2.1.5. In the 2022 Kentucky Derby, a winning \$1 superfecta ticket paid out \$321,500.10 [102]. If you wanted to cover every possible permutation of 4 horses from 20, how much money would you have to invest?

At \$1 per ticket, your total ticket cost would be \$116,280—which is less than the payoff, so this would be a profitable bet *provided* that no one else chose the same 4 horses in the same order. ∎

The challenge to the expert horse bettor is to find information about the horses in the race that suggests that certain outcomes are more or less likely than others—perhaps the favored horse has a history of performing well only in rainy weather, and the forecast for the race is dry. This information can then be used to make bets that are thought to have a better-than-random chance of winning.

Example 2.1.6. If you can safely identify which horse will win, one way to wager is to make a *wheel* bet on the exacta. A wheel combines one horse to win with every one of the other horses to place, and so is guaranteed to win provided that you chose the winner correctly. In a 20-horse race, a wheeled exacta bet is the equivalent of 19 different tickets; a wheeled trifecta bet covers $_{19}P_2 = 342$ different tickets. ∎

Where Order Doesn't Matter: Combinations

Most of the time when gambling, we are not so concerned about the order of events, as when a hand of cards is dealt or a set of lottery numbers is drawn. For counting these arrangements, we are interested in *combinations* rather than permutations.

Definition 2.1.3. A *combination* of r items from a set of n items is a subset of r items chosen without regard to order. The number of such combinations is denoted $\binom{n}{r}$, pronounced "n choose r". An alternate notation is $_nC_r$.

Here, ABC and CBA are interchangeable combinations, as they are subsets of the alphabet consisting of the same three letters. The different order is not a concern here. If the elements of a selected subset are receiving the same treatment once selected, then the choice is a combination, not a permutation.

Example 2.1.7. The American Song Contest, a competition aired on NBC television in 2022, is a musical competition with 56 participants: one from each US state and inhabited territory. If we pick the top four finishers in order, that would be a permutation, since there is a significant difference between finishing first and finishing fourth. If we simply select four of the contestants to appear at a press conference, then it does not matter which singer is selected first, second, third, or fourth—this is a combination. ∎

Theorem 2.1.4. *The number of combinations of r items chosen from a set of n items is*

$$\binom{n}{r} = \frac{n!}{(n-r)! \cdot r!} = \frac{_nP_r}{r!}.$$

Proof. We begin with the formula for the number of permutations:

$$_nP_r = \frac{n!}{(n-r)!}.$$

Since we are looking for combinations, two permutations that differ only in the order of the elements are identical to us. Any combination of r elements from a set of n can be rearranged into $r!$ different orders, by the Fundamental Counting Principle. We then have

$$\binom{n}{r} = \frac{_nP_r}{r!}$$

$$= \frac{n!}{(n-r)! \cdot r!},$$

as desired. □

$\binom{n}{r}$ can be calculated on a TI-84+ calculator by following the steps on page 36 for $_nP_r$ but replacing **2: nPr** by **3: nCr**.

The following theorem collects several simple facts about combinations.

Theorem 2.1.5. *For all $n \geqslant 0$:*

1. $\binom{n}{0} = \binom{n}{n} = 1$, *and* $\binom{n}{1} = n$.

2. For all k, $0 \leqslant k \leqslant n$, $\binom{n}{k} = \binom{n}{n-k}$.

3.

$$\binom{n}{0} + \binom{n}{1} + \binom{n}{2} + \cdots + \binom{n}{n} = 2^n.$$

The left-hand side of this equation can be written more concisely as

$$\sum_{r=0}^{n} \binom{n}{r},$$

where Σ is the capital Greek letter sigma. Sigma corresponds to the letter S and is used here to stand for "sum." The new variable r is an index that keeps track of the terms being added.

Proof.   1. Given a set of n elements, there is only one way to select none of them—that is, there is only one way to do nothing, so $\binom{n}{0} = 1$. Similarly, since the order does not matter, there is only one way to choose all of the items: $\binom{n}{n} = 1$.

If we are choosing only one item, we may select any element from among the n, and there are thus n choices possible.

2. We note that every selection of k items from a set of n partitions the set into two disjoint subsets: one of size k and the other of size $n - k$, and so choosing k items to take is equivalent to choosing $n - k$ items to leave behind. The conclusion follows immediately.

Alternately, direct application of the formula for combinations gives the following:

$$\binom{n}{k} = \frac{n!}{(n-k)! \cdot k!} = \frac{n!}{k! \cdot (n-k)!} = \frac{n!}{[n-(n-k)]! \cdot (n-k)!} = \binom{n}{n-k}.$$

3. If we think of a combination of r items from a set A with $\#(A) = n$ as choosing a subset of A with r elements, then the left side of this equation is simply the total number of subsets of A of all sizes.

From Theorem 2.1.2, we know that the number of subsets of a set with n elements is 2^n. Since we have counted the set of all subsets of A in two different ways, those two expressions must be the same, completing the proof.

$\square$

Horse racing

While exacta, trifecta, and superfecta wagers in horse racing require a gambler to pick the first two, three, or four finishers in order, another wagering option is a *box* bet. Boxing a bet gives the bettor all possible permutations of the selected horses, and thus eliminates the need to pick the exact order correctly. A boxed wager comes with either an increased price or a decreased payoff.

Example 2.1.8. For a 20-horse race, what is the probability of winning a superfecta box wager?

You must select four horses, but the order in which they finish does not matter, which means that your ticket will cover all $_4P_4 = 24$ permutations. Your probability of winning is

$$\frac{_4P_4}{_{20}P_4} = \frac{1}{\binom{20}{4}} = \frac{1}{4845} \approx 2.0640 \times 10^{-4},$$

where the second expression reflects the reality that you are simply trying to pick the right combination of four horses without any regard to order.

The chance of winning has increased over the chance of a straight (un-boxed) superfecta bet, but the payoff will surely be much smaller by comparison. ∎

Poker

When counting possibilities in a poker game like Texas hold'em, we may safely ignore suits, since suits have no standing in hold'em. Considering only the ranks of the cards, there are, of course, $\binom{52}{2} = 1326$ different ways to choose 2 cards from a deck, but this includes many holdings which are indistinguishable: for example, 8♢ 4♠ and 4♢ 8♣. Looking for the number of 2-card sets that are different from a player's perspective, we note that there are 13 ranks for each card, and repeating a rank is allowed—even desirable, as it leads to a pocket pair. The number of possible pocket pairs is 13: the number of ranks in a deck.

Unmatched hole cards are counted by counting ranks. There are

$$\binom{13}{2} = 78$$

ways to select 2 different ranks for hole cards. Taken together with the 13 different pocket pairs, we have 91 different possible hole-card holdings.

Example 2.1.9. The best set of hole cards in Texas hold'em is a pair of aces, often dubbed "pocket rockets" or "bullets" by players. What is the probability of being dealt a pair of aces?

There are $\binom{4}{2} = 6$ different pairs of aces, and $\binom{52}{2} = 1326$ different pairs of cards that can be dealt, since the order does not matter. We find that the probability of pocket aces is

$$p = \frac{6}{1326} = \frac{1}{221}.$$

In the long run, then, you will be dealt a pair of aces once in every 221 hands—which means that if this pair arrives, you should bet it for as long as doing so makes sense, since you have an advantage and this is a genuinely rare event.   ∎

Since the formulas for $_nP_r$ and $\binom{n}{r}$ involve factorials, which are known to grow very fast, it follows that these numbers also get very big very fast. Starting with a standard 52-card deck, the number of possible subsets of a given size r is quite large, even if order is not considered. Three-card poker (page 139) deals a three-card hand to each player. Since the order does not matter, there are

$$\binom{52}{3} = \frac{52!}{49! \cdot 3!} = 22,100$$

different hands. In five-card poker, the number of hands is

$$\binom{52}{5} = 2,598,960,$$

and in contract bridge, where each player receives 13 cards, there are

$$\binom{52}{13} = 635,013,559,600$$

hands—over 600 billion.

Example 2.1.10. In five-card draw poker, the royal flush is the highest-ranked hand. The order in which the cards are dealt does not affect the hand's value—indeed, if you are ever dealt a royal flush, it would be bad form to complain that the cards didn't arrive in increasing order. Since there are four possible royal flushes—the cards are fixed, so the only variable is the suit—the probability of a royal flush is

$$p = \frac{4}{2,598,960} = \frac{1}{649,740}.$$

                                                                                         ∎

In computing the probability of various poker hands, we need to account for every card in the hand—for example, when computing the probability of two pairs, we need to exclude the possibilities that the two pairs are of the same rank, which would mean that the hand was four-of-a-kind, and that the fifth card is the same rank as one of the pairs, which would turn the hand into a full house.

Example 2.1.11. The form of a two-pair hand is $AABBX$, where A, B, and X are card ranks, $A \neq B$, and neither A nor B is equal to X. A and B, together, may be chosen in $\binom{13}{2} = 78$ ways, since the order does not matter. Once the ranks have been chosen, the pairs can consist of any two of the four cards of those ranks, and the order in which we pick them doesn't matter. By the Fundamental Counting Principle, we can find the two pairs in

$$\binom{13}{2} \cdot \binom{4}{2}^2 = 78 \cdot 36 = 2808$$

ways.

The fifth card, X, can be any of the 48 remaining that do not match either pair. For example, if we have a pair of queens and a pair of 7s, the fifth card can be any card remaining except for the two queens and the two 7s, and there are 44 of these.

Putting this all together with the Fundamental Counting Principle and remembering that the number of possible poker hands is $\binom{52}{5} = 2,598,960$, we have

$$P(2 \text{ pairs}) = \frac{2808 \cdot 44}{2,598,960} = \frac{123,552}{2,598,960} \approx .0475 = 4.75\%.$$

∎

In the hierarchy of poker hands (pages 26–27), a flush beats a straight. This is an ordering that people often get backwards, because to the casual observer, getting five cards in sequence but not of the same suit seems very similar to getting five cards of the same suit but not in sequence. By counting the number of possible hands of each type, the reason for this ordering becomes clear.

Flushes are easier to count. We select a suit, in 4 ways, and then choose 5 of the 13 cards in that suit. Multiplication gives us

$$4 \cdot \binom{13}{5} = 5148$$

flushes.

This number includes straight flushes and royal flushes, which a poker player would not play as a flush. To obtain an accurate count, we must subtract these hands from our total. Any card from ace through 9 may be the lowest-ranking card in a straight flush, and so there are 36 of those. For royal flushes, the only variable is the suit, so there 4 royal flushes. Subtracting these 40 hands from 5148 leaves 5108 flushes.

To count straights, while the order in which the cards are dealt doesn't matter, we can facilitate the count by thinking of the cards in numerical order. There are 40 choices for the lowest card in a straight, because an ace can be either high or low, so all cards from ace through ten are candidates. Having chosen this first card, the remaining ranks are determined, and we have four choices for each of the four remaining cards, giving

$$40 \cdot 4^4 = 10,240$$

possible straights. Once again, we need to subtract the 40 straight and royal flushes, leaving 10,200 hands that are called a straight.

Comparing these numbers, we see that there are almost exactly twice as many possible straights as flushes, which accounts for the relative ranking of the two hands.

We consider next some general formulas for counting hands in a 52-card deck [16]. The number of ways to draw exactly h sets of i cards of the same rank and j sets of k cards of the same rank among a hand of m cards, where $hi + jk \leqslant m$, is

$$\binom{13}{h}\binom{4}{i}^h \cdot \binom{13-h}{j}\binom{4}{k}^j \cdot \binom{13-h-j}{m-hi-jk}\binom{4}{1}^{m-hi-jk}.$$

In this formula:

- The factor $\binom{13}{h}\binom{4}{i}^h$ counts the number of ways to choose h ranks from 13 and then to choose i cards from the 4 cards in each suit. The exponent h counts the number of sets of i cards.

- Once this choice has been made, there are $\binom{13-h}{j}\binom{4}{k}^j$ ways to choose j ranks from the $13 - h$ ranks that remain, and then k cards of each chosen rank.

- At this point, there are $13 - h - j$ ranks left untouched, and we want to choose $m - hi - jk$ of them to fill out the hand with singleton cards drawn from the 4 suits.

For example, when counting full houses in a 5-card hand, we have $h = 1, i = 3, j = 1, k = 2$, and $m = 5$, and we have

$$\binom{13}{1}\binom{4}{3}\binom{12}{1}\binom{4}{2} = 3744.$$

full houses.

This formula encompasses the calculations for four of a kind hands, full houses, three-of-a-kinds, two-pair hands, and pairs.

Counting straights and flushes, including straight flushes, involves the following additional formulas:

Straight flushes: $4(15 - m)$

This number includes royal flushes.

$$\text{Flushes:} \quad 4\binom{13}{m} - 4(15 - m)$$

$$\text{Straights:} \quad (15 - m)\binom{4}{1}^{m} - 4(15 - m).$$

The number 15 arises in these formulas because there are 13 ranks in a deck and the ace can rank both high or low in a straight or straight flush. The number of possible lowest cards in a straight is then $13 - (m - 2) = 15 - m$, since the top $m - 2$ ranks, with the ace considered low, are not low enough to fit $m - 1$ cards in sequence above them. If $m = 5$, then $15 - m = 10$, which indicates that any card from the ace through 10 can be the lowest card in a straight or straight flush.

Example 2.1.12. Most Texas hold'em experts advise a fairly conservative betting strategy, folding most initial hands unless they contain a pair, two high cards (such as AK), or certain suited *connectors*, which are two hole cards of consecutive ranks, so called because of their potential to yield a straight. According to this reasoning, which is backed up by millions of simulated hands, a pair of hole cards such as 9♠ 2♠ should be folded immediately.

Some novice players note "But they're *suited!*" and insist on betting that sort of hand. What is the probability of drawing into a flush starting with these two cards?

To get a flush in this example, the flop, turn, and river must contain at least three spades among their five cards. In counting the number of spades in a set of five cards, we need to account for the suits of all five cards, so that a set with four spades is counted with four-spade sets and not also with the three-spade sets. For example, if we wish to count the number of ways to draw exactly three spades, we must count the number of ways to draw three spades, count the number of ways to draw two nonspades, and then use the Fundamental Counting Principle to multiply those two numbers. Given our hole cards, there are 11 spades and 39 other cards remaining to be dealt; it follows that the number of ways to draw five cards containing exactly three spades is

$$\binom{11}{3} \cdot \binom{39}{2}.$$

Accounting for every card, we have

$$\binom{11}{3} \cdot \binom{39}{2} + \binom{11}{4} \cdot \binom{39}{1} + \binom{11}{5} = 135,597$$

ways for the community cards to complete a flush.

The probability of drawing into a flush, therefore, is

$$p = \frac{135,597}{\binom{50}{5}} = \frac{135,597}{2,118,760} \approx .0640 \approx \frac{1}{16},$$

only about a 6.4% chance of success including the very low probability of drawing into a straight flush. Betting on this sort of hand in the hope of drawing a flush is a losing proposition.   ■

There is an exception to this betting vs. folding strategy. Each round of Texas hold'em begins with two designated players making *ante*, or starting, bets before the cards are dealt. These bets, whose amounts are fixed, are called the *small blind* and the *big blind*, which is typically twice the small blind. These bets are often thought of as an initial bet and an initial raise. The players charged with making these bets are identified by a button that rotates clockwise around the table [106].

If it is your turn to make the big blind bet, and no further bets are made in the opening round before the flop, there is no cost to you to remain in the game and see the flop, and thus there is no reason why you should fold any hand, no matter how weak. While the chance of improving two weak hole cards into a winning hand is slight, there is nothing lost in this case in seeing what cards come out on the flop.

Lotteries

An *r/s lottery* is a game where players bet on which *r* numbers will be drawn from a set of *s*, and games operating on this basic premise are common throughout the world. Lotteries are another example of a game of chance where the order of events, in this case the drawing of numbers, does not matter. We are perhaps most familiar with multimillion-dollar lottery prizes that attract media attention, but the typical *r/s* lottery offers prizes other than the life-changing jackpot.

Example 2.1.13. The Michigan State Lottery's Lotto 47 game is a 6/47 lottery, with prize structure shown in Table 2.1.

TABLE 2.1: Michigan Lottery: Lotto 47 pay table

Match	Payoff
3	$5
4	$100
5	$2500
6	Jackpot

The jackpot starts at $1 million and increases each week that it is not won, resetting to $1 million in the drawing after a winner is determined. What are the probabilities of winning each of the four prizes?

The probability of winning each of these amounts is computed using the formula for combinations, recognizing that the order in which the numbers are drawn is not important, only that the numbers you've chosen be drawn somewhere in the set of six. In reporting the winning numbers for a lottery drawing, media outlets will arrange them in ascending order, but this is a convenience and almost certainly not a faithful representation of the order of events. When doing so, it is important to account for each of the six numbers being drawn in both the numerator and denominator.

The probability of matching exactly k numbers out of 6, chosen from a set of 47, is given by the formula

$$P(\text{Match } k) = \frac{\binom{6}{k} \cdot \binom{41}{6-k}}{\binom{47}{6}}.$$

In the numerator, the first term counts the number of ways to choose the winning numbers, while the second is the number of ways to choose the nonwinning numbers from among the 41 not appearing on the ticket. When computing these numbers on a calculator, you should use the built-in $_nC_r$ routine and not round off any results until the end of the calculation. For $k = 3, 4, 5$, and 6, we have the probabilities shown in Table 2.2.

TABLE 2.2: Michigan Lottery: Lotto 47 prize-winning probabilities

k	$P(\text{Match } k)$
3	.0199
4	.0011
5	2.291×10^{-5}
6	9.313×10^{-8}

Summing the probabilities in Table 2.2 shows that the chance of winning *any* prize when buying a Lotto 47 ticket is only .0210, or about 1 chance in 47.5. ∎

More generally, the probability of matching k of the numbers drawn in an r/s lottery is

$$P(\text{Match } k) = \frac{\binom{r}{k} \cdot \binom{s-r}{r-k}}{\binom{s}{r}}.$$

Table 2.3 (page 48) contains a list of international lottery games. The smallest lottery listed there is the Double Daily Grand lottery from the Caribbean island of St. Lucia: a 4/22 lottery admitting only 7315 different combinations. Since the 2022 population of St. Lucia is about 185,000, this lottery is surely adequate for its market. The probabilities of matching k numbers in the St. Lucia lottery are shown below.

TABLE 2.3: A sample of international lottery games

Location	Lottery Name	r	s	$\binom{s}{r}$
St. Lucia	Double Daily Grand	4	22	7,315
Trinidad and Tobago	Cash Pot	5	20	15,504
Guyana	Daily Millions	5	26	65,780
Aruba	Lotto di Dia	5	30	142,506
Dominican Republic	Loto Pool	5	31	169,911
Grenada	Lotto	5	34	278,256
Costa Rica	Lotto	5	35	324,632
Windward Islands	Super 6	6	28	376,740
Iceland	Lotto	5	40	658,008
Poland	MINI Lotto	5	42	850,668
Barbados	Mega 6	6	33	1,107,568
Malta	Super5	5	45	1,221,759
Samoa	Samoa National Lotto	6	35	1,623,160
France	France Loto	5	49	1,906,884
Fiji	Pick 6	6	36	1,947,792
Israel	Lotto	6	37	2,324,784
Jamaica	Lotto	6	39	3,262,623
New Zealand	Lotto	6	40	3,838,380
Mongolia	6/42 Jackpot Lotto	6	42	5,245,786
Norway	Lotto	7	34	5,379,616
Netherlands	Lotto	6	45	8,145,060
Denmark	Saturday Lotto 7/36	7	36	8,347,680
Argentina	Quini 6	6	46	9,366,819
Ireland	Lotto	6	47	10,737,573
Sweden	Viking Lotto	6	48	12,271,512
Germany	Lotto	6	49	13,983,816
India	Lotto India	6	50	15,890,700
Finland	Lotto	7	40	18,643,560
South Africa	Lotto	6	52	20,358,520
Turkey	Süper Loto	6	54	25,827,165
Philippines	Grandlotto 6/55	6	55	28,989,675
Mexico	Melate	6	56	32,468,436
Nigeria	GG 5/90 Chance 5	5	90	43,949,268
United Kingdom	Lotto	6	59	45,057,474
Brazil	Mega-Sena	6	60	50,063,860
Australia	Oz Lotto	7	47	62,891,499
Russia	GosLoto	7	49	85,900,584
Canada	Lotto Max	7	50	99,884,400
Italy	SuperEnalotto	6	90	622,614,630

k	P(Match k)
0	.4183
1	.4462
2	.1255
3	.0098
4	.0001

Double Daily Grand tickets win a prize if they match 2 or more numbers; we see that the probability of holding a winning ticket is approximately .1355.

In New York state, the twice-daily Take 5 drawing is a 5/39 lottery with winning numbers drawn at midday (2:30 P.M.) and in the evening (10:30 P.M.). Tickets must be purchased separately for each drawing. On October 27, 2022, the same combination of winning numbers—18, 21, 30, 35, and 36—was chosen at both drawings [87]. The New York lottery uses numbered ping-pong balls in an air blower, and independent auditors quickly moved to inspect the equipment and confirm that the two drawings were independent. The probability of any combination being repeated at two successive drawings is simply

$$\frac{1}{\binom{39}{5}} = \frac{1}{575,757},$$

since once the first combination is drawn, all we are looking for is the probability that the second drawing matches it. The probability that a *specified* 5-number combination is drawn twice in a row is much smaller:

$$\left(\frac{1}{575,757}\right)^2 = \frac{1}{331,496,123,049}.$$

There were no winning tickets for the midday drawing, but 52 tickets were sold for the winning combination in the evening drawing. Each winner collected $715.50 from the jackpot pool; a lone winner would have received $37,206 [87].

While this may seem like an unlikely, even suspicious, occurrence, a moment's reflection suggests that suspicion may be uncalled for. When we consider the dozens of state, provincial, and national lotteries offering r/s games, the number of different games offered—New York, for example, has 7 different draw games, with varying values of r and s, conducting a total of 28 drawings every week—and the number of years that lotteries have been running these games, it is far less surprising that some draw game somewhere in the world, at some point in time, drew the same combination in 2 consecutive draws.

Another lottery coincidence that initially seems unbelievable happened to Massachusetts resident Maureen Wilcox in 1980. Wilcox purchased lottery tickets in both Massachusetts and Rhode Island. Both of her tickets matched the winning numbers—but her Massachusetts ticket matched the Rhode Island numbers and her Rhode Island ticket matched the Massachusetts numbers [48].

The probability of this event is tiny; the same as the probability that both of her tickets won in the correct state. Once again, given enough gamblers buying enough lottery tickets, even a coincidence this striking is bound to happen to someone at some time.

Powerball is a popular American game administered by the Multi-State Lottery Association. Tickets for this lottery, which cost $2, are sold in 45 states, the District of Columbia, Puerto Rico, and the U.S. Virgin Islands. The game operators choose five white balls from a bin containing balls numbered from 1 to 69 and one red ball from a separate bin with balls numbered from 1 to 26—this last ball is known as the Powerball. Thus, we may regard Powerball as a 5/69 lottery together with a simultaneous 1/26 game. The jackpot is won by a player holding a ticket whose six numbers match those chosen by the lottery operators. Order doesn't matter when choosing the white balls, so they may be chosen in

$$\binom{69}{5} = 11,238,513$$

different ways. By the Fundamental Counting Principle, we must multiply this number by the 26 choices for the Powerball; this gives 292,201,338 possible results for each drawing, and so the probability of winning the jackpot on a single ticket is $\dfrac{1}{292,201,338}$.

To put this number in perspective, consider Michigan Stadium at the University of Michigan-Ann Arbor (Figure 2.1), which is the largest stadium in the USA, with a capacity of 107,601. If we were to fill the stadium 2715 times,

FIGURE 2.1: Michigan Stadium, Ann Arbor, MI [52].

we would have almost as many people as there are possible Powerball tickets. In the history of the stadium, which includes many years before its capacity was expanded to its current number, there have not been even half that many games. In short, fewer people have attended Michigan home football games than there are Powerball combinations.

Powerball's pay table is shown in Table 2.4. In this table, "+ P" denotes a winning combination that includes matching the Powerball.

TABLE 2.4: Powerball pay table

Match	Payoff
5 + P	Jackpot
5	$1,000,000
4 + P	$50,000
4	$100
3 + P	$100
3	$7
2 + P	$7
1 + P	$4
0 + P	$4

Example 2.1.14. Table 2.4 shows that the same payoff, $100, goes to a ticket matching 4 numbers without the Powerball or to one matching 3 numbers plus the Powerball. Are these outcomes close to equally likely?

Matching 4 numbers without the Powerball has probability

$$\frac{\binom{5}{4} \cdot \binom{64}{1}}{\binom{69}{5}} \cdot \frac{25}{26} \approx 2.7378 \times 10^{-5}.$$

The chance of matching 3 numbers and the Powerball is

$$\frac{\binom{5}{3} \cdot \binom{64}{2}}{\binom{69}{5}} \cdot \frac{1}{26} \approx 6.8994 \times 10^{-5},$$

which is $\frac{63}{25} = 2.52$ times more likely than matching 4 numbers without the Powerball. ∎

Powerball officials have changed the game parameters over its existence, often to increase the number of combinations in the hopes of creating bigger jackpots. Table 2.5 shows the change in the numbers of white and red balls since the game began in 1992.

TABLE 2.5: Powerball combination history [89]

Years	White balls (pick 5)	Red balls (pick 1)	Combinations
1992–1997	45	45	54,979,155
1997–2002	49	42	80,089,128
2002–2005	53	42	120,526,770
2005–2009	55	42	146,107,962
2009–2012	59	39	195,249,054
2012–2015	59	35	175,223,510
2015–	69	26	292,201,338

The probability of winning the jackpot—already quite low—decreased by a factor of more than 5 between 1992 and 2022. This had the intended effect: the Powerball jackpot for the drawing of November 7, 2022 set a new record by topping $2 billion. One ticket, sold in Calfiornia, matched all 6 drawn numbers and qualified for the jackpot.

Keno

The same mathematical principles underlying lotteries can be used in a casino to analyze *keno*. Keno is one of the easiest casino games to understand and to play. In a simple keno game, the player chooses anywhere from 1 to 20 numbers in the range from 1 to 80. The casino then selects 20 numbers from this set. Originally this was done with a cage or blower containing numbered balls—called a *goose*—but many casinos now use a computer to generate the numbers. Figure 2.2 shows a modern keno goose.

The player wins if a certain number of the numbers he or she chose appears among the casino's selection. In many casino resorts, the keno games run continuously, and the numbers can be found on video monitors throughout the premises or on the hotel's closed-circuit television channels, so players can monitor their wagers from the restaurants or their hotel rooms. Some casinos broadcast their drawings over the Internet so the games can be tracked by players even when they aren't in the casino. Two Web sites that aggregate and broadcast the keno drawings from multiple casinos are kenocloud.com and kenousa.com. Since some keno operations allow bets on 1000 consecutive games that can take over three days to complete, this is convenient for people who place such a wager.

American casinos have shifted their game offerings away from live keno in recent years; this is perhaps in part due to the cost of maintaining a keno operation. As we shall see, keno is one of the worst casino games for gamblers, who may have moved on to other games as they realize the steep disadvantage they face. At the same time, keno games have found a new place in state

FIGURE 2.2: Keno goose, after 20 numbers have been drawn [84].

lotteries, with Wyoming the latest to add keno to its lottery game lineup, in 2022. State lottery keno games are typically drawn at a central location and the numbers broadcast to participating bars and other sales points, making the overhead cost lower than for a casino relying on multiple people to staff its lightly-patronized game. This model is also successful in Nebraska, where traditional 80-ball keno is regulated by local governments and a wide variety of games may be found across the state.

There are

$$\binom{80}{20} = 3,535,316,142,212,174,320$$

ways for the casino to draw its 20 numbers. This will be the denominator in our keno calculations. In the history of keno, there is no record of any player who has matched all 20 numbers drawn by the casino on a single ticket, an occurrence which has probability

$$\frac{1}{3,535,316,142,212,174,320} \approx 2.829 \times 10^{-19}.$$

It is more likely that you would toss a coin 61 times and get tails on every toss, an event with probability $p = 2^{-61} \approx 4.337 \times 10^{-19}$. If we imagine one keno drawing every second since the Big Bang, approximately 15 billion years ago, this would only account for 4.7335×10^{17} combinations, and the probability of any single combination being drawn in that time is only about 13.4%.

Possibly in recognition of this mathematical fact, the FireKeepers Casino in Battle Creek, Michigan offers a 20-number keno game which, for a $2 bet, pays off the same $100,000 if the ticket matches 15 or more numbers. Since the probability of hitting 15 numbers out of 20 is small (see Exercise 2.14), and the chances of hitting 16 through 20 are smaller still, this is no great advantage to the player.

Casinos who maintain a live keno operation offer a wide range of keno bets. For example, the players' guide for keno at FireKeepers includes 14 different games, and several others are available at electronic keno terminals. Table 2.6 shows the pay table for a $1 bet on their "Mark 7 Numbers" game. If a player

TABLE 2.6: FireKeepers Casino Mark 7 keno pay table

Match	Payoff
4	$1
5	$20
6	$350
7	$10,000

bets more than $1 (the maximum bet is $5), these payoffs are increased by multiplying each payoff by the amount wagered.

Example 2.1.15. What ·is the probability of matching four out of seven numbers and qualifying for the lowest prize—which is a simple refund of the ticket price?

There are $\binom{7}{4} = 35$ ways to select the four numbers that are matched by the casino. We must also account for the other 16 numbers, and they must be chosen from the 73 numbers that the player did not select. This can be done in

$$\binom{73}{16} = 5,271,759,063,474,612$$

ways. By the Fundamental Counting Principle, any subset of 4 matches can be combined with any subset of 16 non-matches to form a winning combination, and so we multiply these two numbers together to find the numerator: there are 184,511,567,221,611,420 (over 184 quadrillion) ways to win this bet. That sounds encouraging until we compare it to the denominator, which is 3,535,316,142,212,174,320—over 3.5 quintillion, and nearly 20 times larger.

$$P(\text{Match } 4) = \frac{\binom{7}{4} \cdot \binom{73}{16}}{\binom{80}{20}}$$

$$= \frac{184,511,567,221,611,420}{3,535,316,142,212,174,320} \approx .0522.$$

The probabilities of the other winning events in the table can be calculated similarly, using the general formula

$$P(\text{Match } k) = \frac{\binom{7}{k} \cdot \binom{73}{20-k}}{\binom{80}{20}}.$$

For the remaining money-winning combinations, this formula gives the following probabilities:

$$P(\text{Match } 5) = \frac{\binom{7}{5} \cdot \binom{73}{15}}{\binom{80}{20}} \approx .0086.$$

$$P(\text{Match } 6) = \frac{\binom{7}{6} \cdot \binom{73}{14}}{\binom{80}{20}} \approx .00073.$$

$$P(\text{Match } 7) = \frac{\binom{7}{7} \cdot \binom{73}{13}}{\binom{80}{20}} \approx 2.440 \times 10^{-5}.$$

Example 2.1.16. Making only a single n-number bet is not terribly interesting for a gambler, nor is it lucrative for the casino. By making a *way* bet, it is possible to cover more combinations of numbers, frequently at a reduced price per wager, and increase the probability of drawing into a winning combination. To make a way bet, the bettor selects several groups of numbers, as in Figure 2.3, where three groups of four numbers, indicated by shaded blocks, have been chosen.

It should be noted that the numbers comprising a block need not be adjacent as shown here; by placing an X through the desired numbers and circling the entire set, any group of numbers may be combined into a block. The player may then specify how the blocks are to be combined. For the ticket shown in Figure 2.3, there are three 4-spot bets consisting of the three individual blocks, three 8-spot tickets, which arise from combining every pair of 4-spot blocks into a "virtual" 8-spot block, and a single 12-spot ticket encompassing all of the selected numbers. If all seven ways were then played, this ticket would be called a 3/4, 3/8, 1/12 seven-way keno ticket. ■

At FireKeepers, each way may be purchased for as little as 5¢, provided that the ticket as a whole covers at least 100 ways, or $5 of bets. A way ticket provides for the possibility of multiple wins, but this possibility is purchased with a greater cost for the ticket as a whole.

1	2	3	4	5	6	7	8	9	10
11	12	13	14	15	16	17	18	19	20
21	22	23	24	25	26	27	28	29	30
31	32	33	34	35	36	37	38	39	40
41	42	43	44	45	46	47	48	49	50
51	52	53	54	55	56	57	58	59	60
61	62	63	64	65	66	67	68	69	70
71	72	73	74	75	76	77	78	79	80

FIGURE 2.3: Keno bet slip with shaded 3/4, 3/8, 1/12 way bet.

Example 2.1.17. The blocks on a way bet need not be the same size. Consider the ticket shown in Figure 2.4, which has seven blocks: three 2-spots and four 4-spots.

1	2	3	4	5	6	7	8	9	10
11	12	13	14	15	16	17	18	19	20
21	22	23	24	25	26	27	28	29	30
31	32	33	34	35	36	37	38	39	40
41	42	43	44	45	46	47	48	49	50
51	52	53	54	55	56	57	58	59	60
61	62	63	64	65	66	67	68	69	70
71	72	73	74	75	76	77	78	79	80

FIGURE 2.4: Keno bet slip with shaded 3/2, 7/4, 13/6, 18/8, 22/10 way bet.

Suppose now that the 20 numbers drawn against this ticket are the following, with the covered numbers bolded:

$$6, \mathbf{7}, 15, 19, 21, 23, \mathbf{24}, 26, 28, 39,$$
$$\mathbf{44}, 45, 48, 49, \mathbf{51}, 64, \mathbf{67}, 69, 71, 73$$

How many wagers are formed by the selected blocks, and what is the net payoff on this way ticket?

These blocks may be combined in the following ways to form virtual tickets with ten or fewer spots:

- The four 2-spots may be played individually.

- There are seven 4-spot tickets: the four 4-spots played individually and $\binom{3}{2} = 3$ arising from all possible combinations of the three 2-spot blocks: {37,38,51,61}, {37,38,67,77}, and {51,61,67,77}.

- Each 4-spot block may be combined with each of the 2-spot blocks to form a separate 6-spot block, and the three 2-spot blocks may be combined for a virtual 6-spot ticket, making thirteen 6-spot choices.

- There are eighteen 8-spot blocks:

  - $\binom{4}{2} = 6$ from combining two 4-spot blocks.

  - $\binom{3}{2} \cdot 4 = 12$ from combining any two 2-spot blocks with a 4-spot block.

- Finally, there are twenty-two 10-spot blocks:

  - Choose any two 4-spot blocks, which may be done in $\binom{4}{2} = 6$ ways, and add in any of the three 2-spot blocks, for a total of 18.

  - Use all three of the 2-spot blocks with any one 4-spot block, totaling four more.

This results in 63 different combinations of the seven blocks. Suppose that you wager the ticket in Figure 2.4 at this 10¢ minimum, which represents a total investment of $6.30. The payoff table, which is shown in Table 2.7, is taken from the casino's $1 wager pay table; payoffs should be divided by 10 to find the amounts won for a 10¢ bet..

TABLE 2.7: Paytable for a $1 keno wager, FireKeepers Casino

	Number of spots marked				
Match	2	4	6	8	10
0					$1
2	$12	$1			
3		$2	$1		
4		$160	$2		
5			$75	$11	$1
6			$2500	$91	$20
7				$1500	$160
8				$15,000	$1000
9					$5000
10					$50,000

The $1500 payoff for matching 7 out of 8 spots is important, as it represents a net win of $1499. This value has been specifically chosen because a net keno payoff of $1500 or more must be reported by the casino to the Internal Revenue Service. (All gambling winnings are judged by the IRS to be taxable income, whether reported directly by the casino or not. Gambling losses, if carefully documented, may be declared as a deduction, but only up to the extent of your winnings.) It is easy to see that multiple wins at the lower levels, where they are most likely, may be necessary to win back the $6.30 cost of the original ticket.

The only winning combinations are from the 4-spot 14, 24, 34, 44, and the combined 2-spots 51, 61, 67, 77, which each return $1 of the wager. The net return on this ticket in this drawing is –$4.30. ■

Example 2.1.18. Perhaps the ultimate way bet ticket is a *king* ticket. A single number played as part of a way bet is called a *king*, and a keno ticket consisting entirely of kings provides a multitude of betting combinations. A ticket with seven kings circled and all combinations activated has 127 possible winning ways. This number is $2^7 - 1$, and may be easily calculated without resorting to a list by referring to Theorem 2.1.2: There are 2^7 ways to choose a subset of the seven kings, but since one of these, the empty set, does not correspond to a keno combination, we subtract 1 from this number.

If all 7 numbers are drawn, an event with probability

$$p = \frac{\binom{73}{13}}{\binom{80}{20}} \approx \frac{1}{40.979},$$

the $12.70 ticket will pay off 127 times, for a total of $5381.30. ■

Example 2.1.19. FireKeepers also offers a keno bet called the "Ring of Fire" bet. This bet allows you to bet on the 32 edge numbers on the betting slip, which are shaded in Figure 2.5.

1	2	3	4	5	6	7	8	9	10
11	12	13	14	15	16	17	18	19	20
21	22	23	24	25	26	27	28	29	30
31	32	33	34	35	36	37	38	39	40
41	42	43	44	45	46	47	48	49	50
51	52	53	54	55	56	57	58	59	60
61	62	63	64	65	66	67	68	69	70
71	72	73	74	75	76	77	78	79	80

FIGURE 2.5: Keno bet slip with Ring of Fire bet shaded.

Among the 18 payoffs, you win $25,000 if none of the edge numbers is selected. What is the probability of this happening?

The calculation is actually quite simple: You win if the 20 numbers chosen by the casino are among the 48 that are *not* on the edge of the bet slip. The probability of this is

$$P = \frac{\dbinom{48}{20}}{\dbinom{80}{20}} \approx 4.73 \times 10^{-6} \approx \frac{1}{211,244}.$$

■

The Football Pools

In Great Britain, a popular gambling pastime for decades was football [soccer] pools, which celebrate their centennial in 2023. Though the rise of the National Lottery cut sharply into their popularity, the Pools were for many years an exciting game of chance which required no actual knowledge of the sport. The leading Pools manager for many years was Littlewoods, which dates back to 1923 and is now part of The Football Pools [46].

Mathematically, the Pools are very much like an r/s lottery. Bettors select 10–12 games from the list of the weekend's 49 designated games, attempting— if skill plays a part in their choices—to pick the games that will end in draws. Tied games are divided into two subsets: *score draws*, where each team scores at least once, and *scoreless draws*, where the final score is 0-0. The Classic Pools game covers all possible combinations, or "lines," of 8 games from those selected. A 10-game ticket costs only £1, and thus covers $\dbinom{10}{8} = 45$ possible combinations at a cost of less than 3 pence apiece. Players may select 11 matches for £2.75, covering $\dbinom{11}{8} = 165$ lines, and 12 games, totaling $\dbinom{12}{8} = 495$ lines, for £7.50 [75].

Points are awarded based on the outcomes each set of 8 games, and a player's ticket is valued at the best score of all 8-game subsets that can be formed. A game ending in a score draw counts for 3 points, a scoreless draw or void match (one where the match is played on a different date than originally declared, for example) for 2, and a win for 1. Prize funds established from entry fees are divided among bettors whose lines score the top three point totals; if 8 games of the 49 end in score draws, the top prize pool is divided among any players with a line totaling 24 points. The top prize fund typically runs into millions of pounds.

Example 2.1.20. Assuming that exactly eight score draws occur, a player buying a 10-game ticket has probability

$$p = \frac{\dbinom{10}{8}}{\dbinom{49}{8}} = \frac{45}{450,978,066} \approx 9.978 \times 10^{-8}$$

of winning a share of the top prize pool.   ■

Many Pools players simply pick their favorite numbers in the range from 1 to 49, thus making this at least as much a game of chance as a game of skill. The Pools are so popular in Britain that in 1963, on a weekend when unusually bad weather led to the cancellation of most games, a "Pools Panel" was appointed to invent outcomes for games so as not to interfere with gambling [46].

2.2 Odds

In the world of gambling, probabilities are often encountered in terms of *odds*.

Definition 2.2.1. The *odds against* an event A is the ratio $P(A^C) : P(A)$, or $P(A^C)/P(A)$.

This is often stated in a form like "x to 1," as is the case in horse racing, for example. Most of the time when odds are quoted, they are odds against. We can also consider the odds in favor of an event.

Definition 2.2.2. The *odds for* an event A is the ratio $P(A) : P(A^C)$—the reciprocal of the odds against A.

Example 2.2.1. When rolling 2d6, *snake eyes* is the name given to a roll of 1-1. Since there are 36 ways for the dice to land, and only one is a 1-1, the probability of snake eyes is 1/36. The odds against snake eyes would then be

$$\frac{35}{36} : \frac{1}{36} = \frac{\frac{35}{36}}{\frac{1}{36}} = \frac{35}{1},$$

or 35 to 1. ∎

Example 2.2.2. The odds against the double street bet (six numbers) at American roulette are

$$\frac{32}{38} : \frac{6}{38},$$

or 32 to 6, which reduces to $5\frac{1}{3}$ to 1. ∎

Note that the payoff for the double street bet is 5 to 1, so this bet is paid off at less than its true odds of $5\frac{1}{3}$ to 1, which is where the casino gets its advantage. Contrary to common belief, a casino doesn't make its money from wagers collected from losing bettors, but rather from paying off winners at less than true odds. Roulette and craps tables, especially, ensure a casino's profits by retaining some money to which the winner of a fair bet paid at true odds would be entitled.

Example 2.2.3. At the 1973 Belmont Stakes, the final race in thoroughbred racing's Triple Crown, the quoted odds on the favored horse, Secretariat, were 1 to 10. Using this information, we can compute $P(A)$, the probability that the oddsmakers assigned to Secretariat winning:

$$\frac{1}{10} = \frac{P(A^C)}{P(A)} = \frac{1 - P(A)}{P(A)}.$$

Cross-multiplying gives

$$P(A) = 10 - 10 \cdot P(A)$$
$$11 \cdot P(A) = 10$$
$$P(A) = \frac{10}{11} \approx .909 = 90.9\%.$$

■

Secretariat won the race by 31 lengths, becoming the first Triple Crown winner since Count Fleet in 1948. Both the margin of victory and the time, 2:24.00, remain Belmont Stakes records.

We can rearrange the formula for odds and derive the following result.

Theorem 2.2.1. *If the odds against an event A are x to 1, then $P(A) = \frac{1}{x+1}$.*

Proof. We have

$$\frac{1 - P(A)}{P(A)} = \frac{x}{1}.$$

Cross-multiplying gives

$$1 - P(A) = x \cdot P(A),$$

or

$$(x + 1) \cdot P(A) = 1.$$

Dividing by $x + 1$ gives

$$P(A) = \frac{1}{x+1},$$

completing the proof. □

More generally, if the odds against an event A are x to y, the probability of the event is $P(A) = \frac{y}{x+y}$.

Applying this result in a casino sports book can lead to some interesting revelations. In December 2022, the Superbook at the Westgate Casino in Las Vegas set odds of winning the championship on the four participants in the

College Football Playoff. These odds, and the corresponding probabilities of winning the championship, are shown in Table 2.8.

TABLE 2.8: Westgate Casino: College Football Playoff odds of winning, December 2022

School	Odds	$P(\text{Win})$
Georgia	5–6	$6/11$
Ohio State	15–4	$4/19$
Michigan	3–1	$1/4$
Texas Christian	33–2	$2/35$

If we were to place a $300 bet (to avoid payoffs rounded down in the casino's favor) on each of the 4 teams, we would invest $1200. If Georgia won, we would receive $550, a net loss of $650. If Michigan won, we would break even, since our $300 bet would be returned with our $900 win. We would only make a profit if one of the two longest shots in the field won.

According to Axiom 2, we would expect the sum of the 4 probabilities to be 1; it was instead

$$\frac{31,107}{29,260} \approx 1.0631.$$

How are we to interpret this deviation from the laws of probability? The extra .0631 reflects the amount of profit that the casino expects to make on futures bets on the playoffs. Westgate has declared its best estimate of the probability of each team winning the championship in setting its odds. If gamblers bet on each team in proportion to the odds assigned by Westgate, then $\frac{1}{4}$ of the wagered money ($25 out of each $100) will be bet on Michigan and $\frac{2}{35}$ ($5.71 of every $100) will be bet on Texas Christian, for example.

The *payoff* on one of these bets will be approximately $100, regardless of which team wins. (Payoffs are rounded down to the nearest cent or 10¢, always in the casino's favor.) Since the 4 probabilities add up to more than 100%, the casino will take in more than $100 for every $100 it eventually pays out to the winners. Specifically, the casino expects to collect $106.31 for every $100 in winnings it will pay.

This excess is the fraction

$$\frac{.0631}{1.0631} \approx .0594,$$

which denotes the "over-round," or proportion of the total money wagered that will *not* be returned to winning gamblers but will be retained by the casino as profit [23]. Here, the casino expects to hold about 6% of the total amount bet.

Westgate was on sound financial footing in offering these wagers. Georgia won the championship, defeating Texas Christian 65–7 in the title game.

2.3 Addition Rules

Our next challenge will be to extend our understanding of probability to *compound* events: events that can be broken down into several simple events. We can often find the probability of these simple events using techniques previously covered; this chapter allows us to combine those probabilities correctly to find probabilities of more complicated events.

Definition 2.3.1. Two events A and B are *mutually exclusive* if they have no elements in common—that is, if they cannot occur together.

Example 2.3.1. If we draw one card from a deck and record its suit, the events $A = \{$The card is a diamond$\}$ and $B = \{$The card is a spade$\}$ are mutually exclusive. ■

Example 2.3.2. Again drawing one card from a deck, the events $A = \{$The card is a diamond$\}$ and $C = \{$The card is a 7$\}$ are not mutually exclusive as there is one card, the $7\diamondsuit$, common to both events. ■

In computing probabilities, we may be in a situation where we know $P(A)$ and $P(B)$ and want to know the probability that either A or B occurs. It is common in mathematics to interpret "or" as the *inclusive* or, which allows for the possibility that both events occur together—"or" here means "and/or". The addition rules described next allow us to compute this new probability in terms of the known ones. There are two rules, depending on whether or not the events under consideration are mutually exclusive.

Theorem 2.3.1. *(The First Addition Rule)* *If A and B are mutually exclusive events, then*

$$P(A \text{ or } B) = P(A) + P(B).$$

The First Addition Rule is, of course, just Axiom 3 with $n = 2$. If A and B are not mutually exclusive, a slightly more complicated formula can be used to calculate $P(A \text{ or } B)$.

Theorem 2.3.2. *(The Second Addition Rule)* *If A and B are any two events, then*

$$P(A \text{ or } B) = P(A) + P(B) - P(A \text{ and } B).$$

Proof. By definition,

$$P(A \text{ or } B) = \frac{\#(A \text{ or } B)}{\#(\mathbf{S})}.$$

What we need to do is compute $\#(A \text{ or } B)$. Elements of A or B can be counted by adding together the number of elements of A and of B, but if any elements belong to both, they have just been counted twice. In order that each element is only counted once, we must subtract out the number of elements that belong to both A and B. This gives

$$\#(A \text{ or } B) = \#(A) + \#(B) - \#(A \text{ and } B).$$

Dividing by $\#(\mathbf{S})$ completes the proof.   □

We can see that the First Addition Rule is a special case of the Second, for if A and B are mutually exclusive, then they cannot occur together; hence $P(A \text{ and } B) = 0$.

Example 2.3.3. Suppose that you are playing five-card draw poker and have been dealt the following hand:

$$4\spadesuit \ 5\spadesuit \ 6\spadesuit \ 7\spadesuit \ K\heartsuit.$$

If you discard the king, what is the probability of your drawing a card that will complete either the straight or the flush?

The question asks for your chance of drawing either a spade, a 3, or an 8. Define the following events:

$$F = \{\text{Draw a spade}\}$$
$$S = \{\text{Draw a 3 or an 8}\}$$

The event F completes a flush, while the event S completes a straight. The intersection of these two events is F and $S = \{3\spadesuit, 8\spadesuit\}$, and consists of the cards that will complete a straight flush.

There are 47 unknown cards remaining in the deck. Of these, nine are spades and eight are either 3s or 8s. It follows that $P(F) = \dfrac{9}{47}, P(S) = \dfrac{8}{47}$, and $P(F \text{ and } S) = \dfrac{2}{47}$, and thus

$$P(F \text{ and } S) = \frac{9}{47} + \frac{8}{47} - \frac{2}{47} = \frac{15}{47} \approx .3191,$$

so you have slightly more than a 30% chance of drawing to complete a straight or flush.   ■

2.4 Multiplication Rules and Conditional Probability

Definition 2.4.1. Two events A and B are *independent* if the occurrence of one has no effect on the occurrence of the other one.

Example 2.4.1. Consider a simple experiment where you toss a fair coin and simultaneously roll a fair d6. Unless you're deliberately trying to bounce one object off the other, the two throws can reasonably be said not to have any influence on each other. We conclude that the results of the toss and the roll are independent events. ∎

Two events that are mutually exclusive (Section 2.3) are explicitly *not* independent, since the occurrence of one eliminates the chance of the other occurring. Moreover, two events that are independent cannot be mutually exclusive.

It is a fundamental principle of gambling mathematics that *successive trials of random experiments are independent.* This includes successive die rolls at craps, successive wheel spins at roulette, and successive weekly drawings of six lottery numbers, but *not* successive hands in blackjack—for in blackjack, a card played in one hand is a card that cannot be played in the next hand. Since the composition of the deck has changed, we are not considering successive trials of the same random experiment.

This principle is not always well-understood by gamblers, and the inability or unwillingness to understand this doctrine of independent trials is sometimes called the *Gambler's Fallacy.* This fallacy is commonly committed by roulette players who have too strong a belief in the Law of Large Numbers, although it can crop up in any casino game where successive trials are independent.

A frequently observed form of the Gambler's Fallacy comes when roulette players rush to bet on red after a long string—three or four—of successive black numbers has come up, on the grounds that red is "due" so that things will "even up." Dice, cards, and roulette wheels don't understand the laws of probability. They have no knowledge of the mathematics we humans have devised to describe their actions, and they certainly don't understand what the long-term distribution of results is supposed to be. For the same reason, it would be equally erroneous to flock to black on the grounds that black is "hot."

Casinos are willing to cater to these mistaken beliefs. At many casinos, the roulette tables are equipped with lighted signs that automatically detect and display the results of the last 20 or so spins. Although the previous spins of the wheel have no effect on the next spin (unless the wheel is defective), if the players wish to track them, identify spurious patterns, and bet accordingly, the casino will gladly accommodate them by providing these results. These signs cost around $7000; one casino installing them found that they paid for themselves in the form of increased roulette income within six weeks [100].

At the Luxor Casino in Las Vegas, the roulette boards at one table displayed the last 14 numbers spun, together with the distribution of red/black and odd/even numbers across the last 100 spins. Individual numbers were labeled as "hot" or "cold" based on the previous 300 spins. Over the course of 300 spins on this triple-zero wheel, any one number would be expected to come up about 7 or 8 times, but it is certain that some numbers, due to random chance alone, would come up more or less often than expected—enough to convince a superstitious observer that certain numbers were favored by the wheel or were overdue to turn up and thus worth backing with a wager.

For example, in a computer simulation of 300 spins, the least-frequent numbers were 000 and 24, each appearing only 4 times. The apparently "hot" numbers, 7 and 25, turned up 13 times apiece.

If A and B are independent events, it is a simple matter to compute the probability that they occur together, with the use of a theorem called the *Multiplication Rule.*

Theorem 2.4.1. *(The Multiplication Rule)* *If A and B are independent events, then*

$$P(A \text{ and } B) = P(A) \cdot P(B).$$

Informally, the Multiplication Rule states that we can find the probability that two successive independent events occur by multiplying the probability of the first by the probability of the second. Mathematical induction can be used to extend this rule to any number of independent events: the probability of a sequence of n independent events is simply the product of the n probabilities of the individual events.

Lucrative though the roulette signs described above may be for the casino, there is no guarantee that the numbers reported on them are accurate, and a careful reading of the fine print on a casino's roulette information brochure will reveal that no guarantee of accuracy is made. This was pointed out in dramatic fashion in June 2012, when an American roulette board at the Rio Casino in Las Vegas showed seven straight 19s. Since the individual spins are independent, the Multiplication Rule can be used to compute probability of getting the same number seven times in a row:

$$p = \left(\frac{1}{38}\right)^6 = \frac{1}{3,010,936,384}.$$

In computing this probability, the first spin can be any number; we are merely computing the probability that the next six spins match it; hence the exponent in the equation above is 6, not 7. While it is within the realm of reason that some roulette wheel somewhere in the world will, given an amount of time less than the average human lifetime, come up on the same number seven times in a row, it was revealed only a couple of days later that the display—though not the wheel itself—had been malfunctioning.

Recall that the payoff on a straight number bet is 35 to 1. Assuming that the display was accurate, had you bet $1 on 19 before the first spin and then let your winnings ride through seven spins—and had the foresight to withdraw your windfall *before* the next spin—you would have won $35 on the first spin, then $36 \cdot 35$ on the second spin, and in general, $35 \cdot 36^{n-1}$ on spin #n. After seven spins, your total winnings would amount to $78,364,164,095: over 78 *billion* dollars. On a practical level, the house limits on maximum bet size would have prevented you from letting your winnings ride more than two or three times. For example, if the casino's limit on inside bets, such as bets on a single number, is $100 (see Figure 6.1 on page 259), you would be unable even to place the third bet in that sequence, which would be for $1295.

Example 2.4.2. We can now use the Multiplication Rule together with the Complement Rule to solve the question posed by the Chevalier de Méré: Why does it seem like the probability of rolling at least one 6 in four tosses of a fair die is slightly more than 50% and the probability of rolling at least one 12 in 24 rolls of 2d6 is slightly less than 50%? In other words, can we develop a theoretical explanation for his observed results?

The probability of rolling at least one 6, by the Complement Rule, is $1 - P(\text{No 6s})$. The probability of *not* rolling a 6 in a single roll is $\frac{5}{6}$. Since the successive rolls of the die are independent, we have

$$1 - P(\text{No 6s}) = 1 - P(\text{Not a 6})^4$$
$$= 1 - \left(\frac{5}{6}\right)^4$$
$$= 1 - \frac{625}{1296}$$
$$= \frac{671}{1296}$$
$$\approx .5177 > \frac{1}{2},$$

consistent with Gombaud's declared experience that he won slightly more than half the time with this wager.

Similarly, the probability of not rolling a 12 in one roll is $\frac{35}{36}$, and the probability of at least one 12 is $1 - P(\text{No 12s})$, or

$$1 - \left(\frac{35}{36}\right)^{24} \approx .4914 < \frac{1}{2},$$

which is also what Gombaud reported. ∎

Note that, by extension, the probability of rolling at least one 6 in n rolls is

$$1 - \left(\frac{5}{6}\right)^n,$$

which approaches but never equals 1 as n approaches ∞.

It is worth noting that the two probabilities in this example are so close to ½ as to cast some doubt on whether Gombaud really gambled at these games long enough to observe the deviation. Since he played a major role in opening up a rich field of mathematics, this discrepancy need not concern us. The story is a good illustration of the difference between theoretical and experimental probability, even if the details may not be completely accurate.

The Birthday Problem

Another use of the Complement Rule and the Multiplication Rule known as the *Birthday Problem* provides an opportunity for placing a bet where you have a clear advantage. If you are in a room with 30 people, offer to bet a companion that at least 2 people in the room have the same birthday—day and month, not necessarily year. Your chance of winning this bet is over 70%.

We begin our analysis by making the following reasonable assumptions:

- No one was born on February 29.

- All other 365 birthdays are equally likely.

- The birthdays of any 2 people are independent of one another. If there are no pairs of twins (or other multiple births) in the room, this is a sound assumption.

Using the Complement Rule, we see that the probability of at least 1 repeated birthday is $1 - P$(No repeated birthdays). Since the birthdays are assumed to be independent, the chance of no repeated birthdays is

$$p = \frac{365}{365} \cdot \frac{364}{365} \cdot \frac{363}{365} \cdots \frac{336}{365}.$$

In this expression, the first person's birthday can be on any day of the year. The next person must be born on one of the 364 other days, so the chance of her birthday being different from the first person's is $\frac{364}{365}$. The third person's birthday can be on any of the 363 days not covered by either of the first 2 people, and so on. Person #30's birthday can be on any of the $365 - 29 = 336$ days not yet covered.

This probability can be rewritten as

$$p = \frac{365 \cdot 364 \cdots 336}{365^{30}} \approx .2937,$$

so the probability of at least one repeated birthday is $1 - p \approx 70.63\%$. In the long run, you should win this bet 7 times out of 10.

This result is counterintuitive; there are 2 reasons why people underestimate this probability.

1. People typically compare the population of the room to the number of birthdays and figure that 30 is far less than 365. It is, but the relevant quantity here is the number of *pairs* of people in the room, which is $\binom{30}{2} = 435$. With so many possible pairs, duplication becomes far more reasonable.

2. Often, people hear the question "What is the chance that two people in this room have the same birthday?" and think "What is the chance that someone in this room *shares my birthday*?" This is a very different question, with a much smaller probability.

The probability that any one person in the room does not share your birthday is $\frac{364}{365}$. Of the 29 other people in the room, the probability that someone shares your birthday is

$$1 - P(\text{No one has your birthday}) = 1 - \left(\frac{364}{365}\right)^{29} \approx .0765,$$

about 1 chance in 13—much less than ½.

For this probability to reach 50% requires 253 other people in the room.

The probability of matching birthdays increases with the number of people in the room. If a 70% chance of winning is not secure enough for you to make this bet, wait until there are 41 people in the room, for then your chance of winning is approximately 9 in 10: 90.32%. If you're content with a mere 50% chance of winning the bet, the chance of at least one shared birthday exceeds 0.5 for the first time when there are 23 people present.

Sports Betting

In the USA, sports betting was legalized nationwide by a Supreme Court decision in 2018, subject to the action of state governments to regulate wagering within their borders. This decision struck down a 1992 law, the Professional and Amateur Sports Protection Act, restricting wagering on sporting events to only four states: Delaware, Montana, Nevada, and Oregon. These four states had all legalized sports betting before the passage of the law.

In the earliest days of sports betting, odds were set on each team playing in a game, with the odds on the favorite shorter than the odds on the underdog. A gambler betting on the favorite might be required to give 2–1 odds, winning $1 for every $2 wagered if the favorite won. By contract, a bettor backing the underdog might receive 8–5 odds, risking $5 for an $8 payoff [39]. This method fell into disfavor because of how it handled lopsided matchups, where gamblers might balk at giving 5–1 or 6–1 odds on an overwhelming favorite or at risking money on a prohibitive underdog with little chance of winning. Gaming changes to adapt to gamblers' interests—if the rules are discouraging action, then the rules can be changed. Sports bets on single games now take two forms: point-spread bets and money line bets.

Point-Spread Betting

If a casino sports book says that the Miami Dolphins are a 3½-point favorite over the Jacksonville Jaguars, bets on Miami only pay off if the Dolphins win by at least 4 points; bets on Jacksonville win if the Jaguars win the game or lose by no more than 3. In effect, a team tabbed as an x-point underdog starts the game—for betting purposes—with x points. Half-point increments are often seen and eliminate the possibility of a wager ending in a tie.

In setting the point spreads for football games, casinos do so in order to encourage approximately equal action on both teams. A 24-point underdog might well get some gamblers willing to back it on those terms: gamblers who wouldn't consider betting on a team that was that overmatched.

The casino makes its money on point-spread bets in the way that bets are accepted: gamblers wishing to bet on a game must usually bet $11 to win $10. This extra $1 is called the *vigorish*, or "vig" for short. The sports book collects $11 from every player and pays out $21 (the original $11 bet plus $10 in winnings) to approximately half of them, leaving the extra $1 vig from each of the losing bettors as the casino's profit: 4.55% of the total handle—and it doesn't matter which team wins.

Of course, point spreads are sometimes known to shift in the days preceding a game, due to changing circumstances such as player injuries or a casino's desire to keep its books balanced with approximately equal action on each team. In our example, if too many people are betting on Miami giving 3½ points, the casino may change the line to "Miami −4½," meaning that the Dolphins are now a 4½-point favorite, with the intent of encouraging people to bet on Jacksonville and balance the amounts being wagered on each team.

This can allow an observant and quick-acting gambler the chance to make separate bets on both teams in a game and hope for a "straddle," where both bets win. For example, if someone has already placed a bet on Jacksonville getting 3½ points above, and the line moves to make Miami only a 2½-point favorite, making a second bet on Miami on these terms means that if the Dolphins win by exactly 3 points, both of these bets pay off.

The risk here is that in the event of any other outcome to the game, one bet wins and the other loses, for a net player loss of the $1 vig, so this may not be advisable if the line has only moved by one point.

Another form of 11–10 betting involving points is available on the *over/under line*: a bet made against a predicted total number of points scored by both teams in the game. If an American football game is listed with an over/under line of 41 points, bettors may choose to be $11 to win $10 that the actual total will be more than or less than this number. If the total score is exactly 41, all bets push, and are refunded. To eliminate the chance of a push, total point bets may also be stated to half a point.

Example 2.4.3. At six casinos in metropolitan Las Vegas operated by Red Rock Resorts, it is possible to wager on multiple sporting events with 1 wager through a daily *parlay card*. In playing a parlay card, you place a bet (minimum bet $2) and pick the winner, against a point spread, of 3–10 games being played on a given day. If *all* of your selections win, you are paid off according to Table 2.9. If one or more bets tie, they are dropped from consideration and the parlay is evaluated at the appropriate lower level. If there are so many ties that fewer than 3 games still have action, the entire remaining ticket has no action and the cost is refunded.

Once again, the payoffs quoted here are in the form "*x for* 1" rather than "*x to* 1," as we saw with the craps table in Figure 1.5, and as is also seen on some video poker and blackjack machines. If you win, your wager is returned

TABLE 2.9: Red Rock Resorts' sports parlay card payoffs

Wins	Payoff
3 for 3	6 for 1
4 for 4	12 for 1
5 for 5	23 for 1
6 for 6	45 for 1
7 for 7	80 for 1
8 for 8	160 for 1
9 for 9	320 for 1
10 for 10	800 for 1

as part of your payoff—for example, if you bet \$10 on a five-team parlay card and pick all five games correctly, you will be paid \$230, not \$240. One reason for quoting odds in this fashion is to make the payoff seem bigger; another reason is that there can be a considerable time interval between when the bet is placed and when the games are completed and the win is verified, and it makes no sense for the casino not to collect your wager when you place the bet.

Since each game can reasonably be considered to be independent of all the others, computing the probability of winning is easy. The casino sets the point spreads in an effort to encourage equal action on each team in each game, so we will begin by assuming you have a 50% chance of picking each winner correctly. With that probability and the assumption of independence, the probability of winning an n-team parlay is simply $\left(\frac{1}{2}\right)^n$. If you are an expert in predicting winners and can pick the right team 55% of the time, your chance of winning rises to $(.55)^n$. ∎

Another parlay card option when betting American football games is a *teaser* bet. Teaser bets give bettors additional points—frequently 6, 6½, or 7—on the team they select, making the team more likely to win against a new point spread. Taking the Detroit Lions +4 on a 6½-point teaser converts the bet into Detroit +10½. Conversely, a favored team might shift from giving points to receiving points, as when a Minnesota Vikings −3 bet becomes Minnesota +4 when played on a 7-point teaser card.

The catch in playing teaser cards is that the payoffs are sharply reduced from ordinary parlay card payoffs: less than even money for 2-team teasers, for example. Table 2.10 shows the pro football pay table at the Golden Nugget in Las Vegas for standard parlay cards and for 6-, 6½-, and 7-point teasers.

Given the change in the point spread, we would expect that your chance of picking each winner would increase. Even a 70% chance of winning a single game, though, would still make your probability of winning all the bets less than 50%.

TABLE 2.10: Golden Nugget pro football teaser payoff odds

Teams	Parlay	6-pt. Teaser	6½-pt. Teaser	7-pt. Teaser
2	13–5	5–7	2–3	5–8
3	6–1	8–5	7–5	11–10
4	11–1	12–5	2–1	8–5
5	22–1	4–1	7–2	3–1
6	40–1	6–1	5–1	9–2
7	75–1	8–1	7–1	6–1
8	140–1	10–1	9–1	8–1

Example 2.4.4. 6 and 7-point teasers are ideally suited to football games. In a casino error that created a massive opportunity for observant bettors, an employee at the Comstock Casino in Reno once posted a teaser card including National Hockey League games with *6 goals* added [3].

Very few NHL games are decided by more than 6 goals, so adding 6 goals to either team's score generated an almost-certain winner on both sides of the wager. The teasers got good action before management shut them down, and the casino took a financial beating. The sports book closed not long afterward, followed shortly by the Comstock itself. ■

In the other direction, a bettor playing a *reverse teaser* or *pleaser* card gives points back to the casino.

Example 2.4.5. At the Circa Casino, the December 18, 2022 Philadelphia Eagles-Chicago Bears game was listed as Philadelphia –9½. On the reverse teaser card, the line shifted by –7 points, so a bettor wishing to include the Eagles took them at –16½ points and could bet on the Bears +2½.

Philadelphia won the game 25–20, so neither side scored a win on a reverse teaser card. ■

In exchange for a smaller chance of winning, the payoffs are bigger. Table 2.11 compares the pay schedules for the pro football teaser and reverse teaser cards played at Circa Casino.

TABLE 2.11: Circa teaser and reverse teaser payoff odds

Teams	Teaser	Reverse Teaser
3	1.6–1	25–1
4	5–2	75–1
5	4–1	250–1
6	6–1	750–1
7	8.5–1	2500–1
8	12–1	7500–1

Money Line Bets

An alternate approach to moving a point spread is offering a *money line*, which requires that bettors wishing to wager on the favorite risk more money than if they were betting on the underdog. In the Miami-Jacksonville game above, rather than projecting a point spread, the bookmaker may set the wager to something like "−120/+110," which means that a gambler wishing to bet on the favorite (Miami) must risk $120 to win $100, while someone betting $100 on the underdog (Jacksonville) would win $110 for a $100 bet. Winning bets require that the selected team win outright, with no margin of victory specified. Moving the line, or changing either or both of the numbers in response to incoming action, substitutes for changing the point spread as a means to encourage approximately equal action on each team. The casino offering money line bets avoids the prospect of being caught in a straddle, because there is no point spread to change.

Both options can be available at the same sports book. In June 2022, the William Hill sports book offered two ways to bet on the United States Football League game between the Tampa Bay Bandits and the Birmingham Stallions. The Stallions were listed as 3½-point favorites, so making a point-spread bet meant giving 11 to 10 odds and either taking the points on Tampa Bay or giving them on Birmingham. The money line was −165 on Birmingham and +145 on Tampa Bay.

Example 2.4.6. Money line bets can be made on propositions other than actual games. In November 2016, right after the NHL's expansion Vegas Golden Knights revealed their nickname, the Westgate Superbook in Las Vegas offered a −50,000 proposition bet that the team would *not* win the Stanley Cup in its first season: 2017–18. A bettor making this wager would have to put up $50,000 about 18 months ahead of the playoffs, and would collect $50,100 unless the Knights won the championship.

On one hand, this seems like a fairly safe bet at plausible odds, since expansion teams are typically not very good in their first season. However, the Golden Knights advanced all the way to the Stanley Cup finals before losing in 5 games to the Washington Capitals, which certainly would have made some big bettors nervous. No one actually made this bet, although it could have been viewed as an oddly reasonable investment opportunity. A winning wager would effectively pay 0.2% interest over 18 months, which was better than banks were then paying on savings.

There was considerable action on the reverse proposition that the team would win the title in their first season, which was a 500–1 wager at the start of the season; gamblers could bet $100 with a chance of winning $50,000 [69]. The odds fell as low as 9–2 over the course of that successful first season. ∎

For comparative purposes, money line bets can be easily converted to odds bets. For a negative money line of $-X$, the wager is X and the payoff is $X + 100$, so the odds in force are effectively $(X + 100)$ to X, or $1 + \dfrac{100}{X}$ to 1.

When betting \$100 on a positive money line of X, the payoff on a winning bet is $X + 100$ and the corresponding odds are $1 + \dfrac{X}{100}$ to 1.

Conditional Probability

If events A and B are not independent, we will need to generalize Theorem 2.4.1 to handle the new situation. This generalization requires the idea of *conditional probability*. We begin with an example.

Example 2.4.7. If we draw one card from a standard deck, the probability that it is a king is $\dfrac{4}{52} = \dfrac{1}{13}$. If, however, we are told that the card is a face card, the probability that it's a king is $\dfrac{4}{12} = \dfrac{1}{3}$—that is, additional information has changed the probability of our event by allowing us to restrict the sample space. If we denote the events "The card is a king" by K and "The card is a face card" by F, this last result is written $P(K|F) = \dfrac{1}{3}$ and read as "the (conditional) probability of K given F is $\dfrac{1}{3}$." ∎

The fundamental idea here is that more information can change probabilities. If we know that event A has occurred and we're interested in event B, we are now not looking for $P(B)$, but $P(B \text{ and } A)$, because only the part of B that overlaps with A is now possible. With that in mind, we have the following formula for conditional probability:

Definition 2.4.2. The *conditional probability* of B given A is

$$P(B|A) = \frac{P(B \text{ and } A)}{P(A)}.$$

This formula divides the probability that both events A and B occur by the probability of the event A that we know has already occurred. Note that if A and B are independent, we immediately have $P(B|A) = P(B)$, since then $P(B \text{ and } A) = P(A) \cdot P(B)$. This is one case where more information—in this case, the knowledge that A has occurred—does not change the probability of B occurring.

Example 2.4.8. Suppose that we draw one card from a standard deck. Let A be the event that the card is an ace, and D be the event that the card is a diamond. We have $P(A) = \dfrac{1}{13}$, and the event "A and D" is the event that the card is the $A\Diamond$. If we are told that the card is a diamond, the probability that the card is also an ace is

$$P(A|D) = \frac{P(A \text{ and } D)}{P(D)} = \frac{1/52}{1/4} = \frac{4}{52} = \frac{1}{13},$$

which confirms that events A and D are independent. Knowing the suit of a drawn card gives us no new information about its rank. ∎

Example 2.4.9. In craps, the probability of rolling an 11 is $\frac{1}{18}$. Suppose that, on the come-out roll, one of the dice leaves the table and cannot be seen, but the other die is a 5. What is the probability that the roll is an 11?

Let A be the event {The first die is a 5} and B be the event {The sum is 11}. We seek $P(B|A)$. $P(A) = \frac{1}{6}$, and $P(B$ and $A) = \frac{1}{36}$, since the compound event B and A can be described as "the roll is a 5 on the first die (the one we can see) and a 6 on the second die." It follows that

$$P(B|A) = \frac{P(B \text{ and } A)}{P(A)} = \frac{1/36}{1/6} = \frac{1}{6}.$$

We see that knowing that one die is a 5 has improved the chance of rolling an 11 from $\frac{1}{18}$ to $\frac{1}{6}$.

In actual casino play, this roll would be declared void, since a die left the table. ■

Example 2.4.10. Recall from Example 2.1.9 that the probability of being dealt pocket aces at Texas hold'em is $\frac{1}{221}$. If you have indeed been dealt two aces, what is the probability that the player to your immediate left also holds two aces in the hole?

We can answer this question by restricting the sample space. With your cards removed from consideration, there are now $\binom{50}{2} = 1225$ possible two-card combinations available for the next player. Of these, only one is a pair of aces (since only two aces remain among the unknown cards), so the probability of a second pair of pocket aces, given that one pair has already been dealt, is

$$P(\text{Your opponent has 2 aces} \mid \text{You have 2 aces}) = \frac{1}{1225}$$

—about five and a half times less, which gives more support to the idea that you should bet your holdings strongly. ■

The Monty Hall Problem

In the early 1990s, the following question, or some variation of it, pushed recreational mathematics and conditional probability into the national spotlight. This problem is a variation of an older puzzle, the Three Prisoners Problem.

> Suppose you're a contestant on a game show, and you're given the choice of three doors. Behind one door, selected at random, is a car; behind the others, goats. You pick a door, say #1, and the host, who knows what's behind the other doors, opens another door which reveals a goat—this is always possible since only one

door conceals a car. He then says to you, "Do you want to trade your door for the other closed door?".

　　　Is it to your advantage to switch?

Frequently, the host was said to be Monty Hall, long-time host of the TV game show *Let's Make A Deal*, and so this came to be known as the *Monty Hall Problem*. There were two popular interpretations of the "correct" course of action, phrased here in terms of how the new information has or hasn't changed your chances of winning.

1. If one door is shown to be a loser, that information changes the probability of either remaining choice—neither of which has any reason to be more likely—to $\frac{1}{2}$. It follows that there is no advantage to switching.

2. Suppose that you pick door #1. The following are the 4 possible outcomes:

   a. The car is behind door #2 $\left(p = \frac{1}{3} \right)$, and Monty opens door #3. You should switch.

   b. The car is behind door #3 $\left(p = \frac{1}{3} \right)$, and Monty opens door #2. You should switch.

   c. The car is behind door #1, and Monty opens door #2 $\left(p = \frac{1}{6} \right)$. You should not switch.

   d. The car is behind door #1, and Monty opens door #3 $\left(p = \frac{1}{6} \right)$. You should not switch.

   It follows that you should switch $\frac{2}{3}$ of the time, and not switch $\frac{1}{3}$ of the time.

Of course, since this is a one-time decision, any strategy is unsuitable for long-term play. If you're interested in the greatest possible chance of winning the car, which reasoning is correct?

The key to resolving this question lies in the fact that the host knows where the car is and will always open a door concealing a goat; his choice is not made randomly. *In effect, you are being offered a choice to take your original door or **both** of the other doors.* If the car is equally likely to be behind any of the 3 doors, then the chance that you originally selected the door with the car is $\frac{1}{3}$ and your chance of winning the car if you switch is $\frac{2}{3}$. Since you know that given any 2 of the 3 doors, at least 1 must have a goat behind it, revealing a goat effectively shifts that $\frac{2}{3}$ probability to the door you are being offered.

Under the reasonable assumption that you would rather win a car than a goat, you should always switch doors. While counterintuitive, numerous computer simulations have experimentally confirmed the theoretical probability that switching doors wins the car $\frac{2}{3}$ of the time.

The General Multiplication Rule

As with the addition rules, we can state a second, more general, version of the Multiplication Rule that applies to any two events—independent or not—and reduces to the first rule when the events are independent. This more general rule simply incorporates the conditional probability of B given A, since we are looking for the probability that both occur.

Theorem 2.4.2. *(The General Multiplication Rule) For any two events A and B, we have*
$$P(A \text{ and } B) = P(A) \cdot P(B|A).$$

Proof. This result follows from the fact that $P(A \text{ and } B) = P(B \text{ and } A)$ and from Definition 2.4.2. □

As an example of the General Multiplication Rule, we consider *bingo*. Bingo is related to lotteries and keno, and is a game familiar to many from its frequent appearances in charitable fundraisers. Foxwoods Casino in Mashantucket, Connecticut, one of North America's largest casinos, got its start as a bingo hall on the Mashantucket Pequot reservation.

Bingo is a simple game to play: players receive cards on which the word "BINGO" is printed above a 5×5 grid of numbers, as in Figure 2.6. The five

FIGURE 2.6: Bingo card.

squares under the B are filled with numbers in the range 1–15, the I squares from 16 to 20, the N squares from 31 to 45, the G squares from 46 to 60, and the O squares from 61 to 75. The center square of the grid, under the N, is traditionally a "free" square and contains no number.

Numbers in the range from 1 to 75 are drawn, either from a cage or blower full of ping-pong balls as in keno, or electronically. As numbers that appear on the players' cards are drawn, the players cover them in some fashion. The free space is considered covered automatically. Depending on the rules of the game, the winner is the first player to complete a line of five covered spaces horizontally, vertically, or diagonally, including the center free space; cover all 24 numbers (the "Cover All" game); or make other significant patterns, such as crosses or Xs, on their card. A winning player traditionally shouts "Bingo!" to indicate their success and have their win certified so that they may claim their prize.

Casinos have been known to offer bingo as a *loss leader*: a game on which they are prepared to lose money in order to bring people into the building. Casino bingo is typically scheduled in discrete sessions, often one starting every other hour, rather than being continuously played. The thought behind a loss leader strategy is that bingo players will venture into other areas of the casino and gamble there—in less favorable games—in between sessions. The reason that bingo may be a losing proposition for a casino is that there is always a winner in every game, no matter how few people might be playing, and if the amount of the top prize is not covered by player wagers, the game still goes on.

Example 2.4.11. How many possible bingo cards are there?

Order matters when laying out a bingo card—if the first five numbers drawn are B-13, I-28, N-41, G-55, and O-72, and all five appear on your card, you are only a winner if they are all in the same row or if all except 41 (the N number) fall on a diagonal. For the B, I, G, and O columns, there are $_{15}P_5 = 360,360$ possible choices of the numbers. In the N column, the free space means that there are only 4 numbers to be picked, and so there are $_{15}P_4 = 32,760$ ways to fill it in. The total number of cards is therefore

$$(360,360)^4 \cdot 32,760 = 552,446,474,061,128,648,601,600,000$$

—or about 552 septillion. ∎

Example 2.4.12. Since the chance of winning at bingo depends both on the numbers on one's own card and the numbers on the other players' cards, it can be challenging to compute the probability of winning. We shall consider a simple question: What is the probability of covering all 24 of your numbers after exactly N numbers have been called?

As with keno, we need to account for all N numbers drawn. Twenty-four of these must be the numbers that appear on your card, and one of them must be the last number; otherwise, you would win in fewer than N numbers. It follows that the first $N - 1$ numbers drawn must contain 23 of yours, and then that, given the occurrence of this event, the last draw must be your 24th number. Define the following events:

- A = Draw 23 numbers in exactly $N-1$ bingo balls.

- B = Draw your 24th number on the Nth ball.

Then $P(\text{Draw 24 numbers in exactly } N \text{ balls})$ is $P(A \text{ and } B) = P(A) \cdot P(B|A)$.

For $P(A)$, there are $\binom{24}{23} = 24$ ways to pick those 23 numbers. The remaining $(N-1) - 23 = N - 24$ numbers must be drawn from among the 51 numbers that are not on your card; those may be chosen in $\binom{51}{N-24}$ ways. There is a total of $\binom{75}{N-1}$ ways to draw the first $N-1$ numbers. Assembling these factors gives

$$P(A) = P(\text{Draw 23 numbers in } N-1 \text{ bingo balls}) = \frac{\binom{24}{23} \cdot \binom{51}{N-24}}{\binom{75}{N-1}}.$$

Having accounted for this event, we now consider the probability of drawing that last number on the Nth bingo ball, given that 23 numbers have already been drawn. When the last number is drawn, there are $75 - (N-1) = 76 - N$ bingo balls remaining, and only one of them bears the lone uncovered number on your card. The probability of matching that number is therefore

$$P(B|A) = \frac{1}{76 - N}.$$

In the N draws, then, we have the following probability:

$$P(\text{Win in exactly } N \text{ draws}) = \frac{\binom{24}{23} \cdot \binom{51}{N-24}}{\binom{75}{N-1}} \cdot \left(\frac{1}{76 - N}\right).$$

∎

We shall illustrate the use of this formula by considering the extreme cases. If $N = 24$, the formula simplifies to

$$P(\text{Win in 24 draws}) = \frac{24}{\binom{75}{23}} \cdot \left(\frac{1}{52}\right) = \frac{24 \cdot 23! \cdot 52!}{75! \cdot 52} = \frac{24! \cdot 51!}{75!} = \frac{1}{\binom{75}{24}},$$

as we would expect, since we're asking for the probability of drawing precisely your 24 numbers in 24 draws.

If $N = 75$, then we have

$$P(\text{Win in 75 draws}) = \frac{24}{75} = .3200,$$

which is just the probability of choosing one of the numbers on your card from the full range of 75 numbers. This may be interpreted as choosing one number *not* to be drawn from 75, and thinking of that number as the last number left in the cage after 74 numbers have been drawn.

In general, the expression for this probability may be rewritten as

$$P(\text{Win in } N \text{ draws}) = \frac{24 \cdot 51! \cdot (N-1)!}{(N-24)! \cdot 75!}.$$

Table 2.12 collects this probability for various values of N.

TABLE 2.12: Bingo probabilities for Cover All

N	P(Cover All in Exactly N Draws)
24	3.8792×10^{-20}
25	9.3100×10^{-19}
30	1.8427×10^{-14}
35	1.1098×10^{-11}
40	1.4629×10^{-9}
45	7.8073×10^{-8}
50	2.2632×10^{-6}
55	4.2125×10^{-5}
60	5.5941×10^{-4}
65	5.6916×10^{-3}
70	4.6664×10^{-2}
75	.3200

At the same time, we might be interested in the probability of covering a complete card in N *or fewer* draws. Computing that probability is a simple matter of adding up the individual mutually exclusive probabilities P(Cover all in x draws) for $24 \leqslant x \leqslant N$:

$$P(\text{Cover all in} \leqslant N \text{ draws}) = \sum_{k=24}^{N} \frac{24 \cdot 51! \cdot (k-1)!}{(k-24)! \cdot 75!}.$$

Some of these cumulative probabilities are collected in Table 2.13.

TABLE 2.13: Cumulative bingo probabilities for Cover All

N	P(Cover All in $\leqslant N$ Draws)
25	9.6979×10^{-19}
35	1.6185×10^{-11}
45	1.4639×10^{-7}
55	9.6537×10^{-5}
65	1.5415×10^{-2}
75	1

In practice, many cards are in play in a single round of bingo: there are surely multiple players, and it is not uncommon for bingo players to play many cards in one game. Calculating the probability of holding the *first* card to win a Cover All game is extremely tricky because two cards are generally not independent; they may share one or more numbers. A rough approximation of the chance of any one card being the first to have all of its numbers called can best be done by repeated simulation—see [119] for the details.

Punchboards

The *punchboard* was a popular, though often crooked, gambling device frequently found in American retail stores as a trade stimulator from the late 1700s through the middle of the 20th century [104]. A common form of punchboard consists of a thick piece of cardboard with anywhere from 20 to 10,000 small holes, called *spots*, cut into it and covered on both sides by paper or thin metal foil. The holes each contain a small slip of paper printed with a number. One such board is shown in Figure 2.7.

FIGURE 2.7: Ninety Percenter punchboard.

Upon payment of an entry fee, which typically ranged between 5¢ and $1, a player would punch out a circle to reveal their number. Certain of these numbers corresponded to prizes of cash or merchandise, which could be redeemed on the spot. Proprietors would purchase punchboards from local dealers for an amount less than the difference between the cost of all of the spots and the sum of all prizes; this is how a merchant could make money from a punchboard in his or her store.

Example 2.4.13. The *Ninety Percenter* punchboard shown in Figure 2.7 claims just that: a 90% payout of all money invested by players. The punchboard has 800 spots, which are sold for 10¢ each. Players punching out any multiple of 10 between 10 and 650 receive a $1 prize, and the last spot on the board carries a $5 bonus. If the claim is accurate and all of the winning numbers appear at least once each, how many multiples of 10 are repeated on this punchboard?

If all 800 spots are sold, the proprietor takes in $80. If the claim of a 90% payback rate is correct, then $72 must be returned in prizes. Subtracting the $5 bonus for the last spot leaves $67, so two extra winning spots must be included—either two numbers repeated twice each or one number repeated three times. ∎

If we make the reasonable assumption that each winning number appears only once, then the total prize money paid out is $70, and the return percentage is 87.5%—which only becomes 90% through generous rounding.

An alternate punchboard game replaced the numbers with girls' names, as in Figure 2.8. Each spot sold corresponded to one name; the person buying each spot signed the punchboard beside their chosen name—perhaps the name of a wife or girlfriend—and then punched out the spot to reveal the price of that name. Once all of the spots were sold, a seal on the punchboard was broken to reveal the prize-winning name.

FIGURE 2.8: Lucky Girl punchboard.

Suppose that you're playing the Lucky Girl punchboard in Figure 2.8, which has 35 spots, and that the top prize is $5. If you are the first to purchase a chance, your probability of winning is $\dfrac{1}{35}$. What is your chance of winning (that is, of picking the winning name) if you are the second player?

To win on the second punch, the first punch must have chosen a losing name. The probability of losing on the first punch is

$$P(A_1) = \frac{34}{35},$$

and the probability of then winning on the second punch is

$$P(A_2|A_1) = \frac{1}{34},$$

since we can compute this conditional probability by restricting the sample space to the 34 spots left after the initial loser. These events must occur in sequence, which gives

$P(\text{Lose on first punch and Win on second punch}) = P(A_1 \text{ and } A_2).$

By the General Multiplication Rule , this is

$$P(A_1 \text{ and } A_2) = P(A_2|A_1) \cdot P(A_1) = \frac{1}{34} \cdot \frac{34}{35} = \frac{1}{35},$$

which is the same as the probability of winning on the first punch. This assumes, of course, that the punchboard operator is offering a fair game, which was often not the case.

Repeating this calculation with different numbers of previous losing punches reveals an interesting and possibly counterintuitive result: *The probability of winning any specified prize is independent of the number of previous punches.* There is no benefit in going first, nor in waiting until the end; the advantages and disadvantages cancel out at any place in the line.

Example 2.4.14. To illustrate this independence, we consider a generalization: For a punchboard with n spots and a single prize, find the probability that the jackpot will be won by the player picking the last spot.

We need only compute the probability that the first $n-1$ spots chosen do not win the jackpot, for then the jackpot number must be under the last remaining spot. The probability of losing on the first punch is

$$p_1 = \frac{n-1}{n}.$$

Given that the first punch loses, the probability that the second punch also loses is

$$p_2 = \frac{n-2}{n-1}.$$

We continue in this fashion:

$$p_k = P(\text{Lose on punch } \#k \mid \text{All previous punches lose}) = \frac{n-k}{n-(k-1)}.$$

Multiplying these probabilities together for $k = 1, 2, \ldots, n - 1$ yields

$$P(\text{Win on punch } \#n) = \frac{n-1}{n} \cdot \frac{n-2}{n-1} \cdot \frac{n-3}{n-2} \cdots \frac{3}{4} \cdot \frac{2}{3} \cdot \frac{1}{2} = \frac{1}{n},$$

since all of the intermediate numerators and denominators cancel.   ∎

This is the same chance of winning that you have if you pick first or second—or anywhere else in line.

Some punchboards, called *keyed* punchboards, were sold together with a list or diagram of the locations of the winning numbers for the use of the proprietor, and while this could easily be explained as a convenience, it was just as easy for an unscrupulous operator to either steer his preferred customers to the winning numbers or punch out all of the biggest winners before putting the board on display so that the top prizes could not be won, thus increasing his profit. Alternately, an associate of the punchboard dealer could obtain a key from him, take it around to neighborhoods where punchboards were available for play, and punch out winning spots in what appeared to be a run of pure luck.

The spread of crooked punchboards led to a rapid decline in their popularity; however, the punchboard features of player involvement in a game and an instant decision make them a forerunner of the scratch-off instant lottery tickets that remain popular today.

Example 2.4.15. Alaska has no state-run lottery and does not participate in Powerball, but Lotto Alaska is authorized to conduct games of chance as charitable fundraisers. *Chase the Ace* is a game where one winning ticket is chosen each week from all tickets sold that week—unlike Powerball, there is exactly one winner per week. The holder of the winning ticket receives 20% of the week's ticket sales and has a chance at winning a jackpot of 30% of sales, with a minimum top prize of $10,000. The remaining 50% of sales is set aside for charitable organizations [73].

The winning ticketholder is invited to choose a number between 1 and 54, which corresponds to a space on the Chase the Ace game board. Each space holds a playing card, including 2 jokers. If the number chosen corresponds to the $A\spadesuit$, then the player wins the jackpot. If not, the number is removed from consideration, and subsequent weeks' winners must choose other numbers. Once the $A\spadesuit$ is selected, the jackpot resets to $10,000, all windows are put back into the pool, and a new game begins.

Figure 2.9 shows the online game board from which weekly winners choose their cards. The jackpot was $95,218 for the drawing on October 30, 2022 [73]. While the chance of holding the winning ticket varies from week to week based on the number of tickets sold, it is simple to show that the probability of drawing the $A\spadesuit$ is independent of when a player wins.

Imagine all 54 weekly winners assembled in a room and ready to choose their space on the board, in order. Just as we saw in Example 2.4.14, the probability of any one winner choosing the $A\spadesuit$ is conditional, involving the

FIGURE 2.9: Lotto Alaska Chase The Ace game board. Used with permission.

probability that no previous weekly winner has chosen that card and the probability of choosing the $A\spadesuit$ from the cards that then remain. The chance of picking the $A\spadesuit$ on the kth card is

$$\frac{1}{54 - (k-1)} = \frac{1}{55 - k}.$$

This probability of winning is then

$$P(\text{Win on card } \#k) = \left(\frac{53}{54} \cdot \frac{52}{53} \cdot \frac{51}{52} \cdots \frac{55-k}{56-k}\right) \cdot \frac{1}{55-k} = \frac{1}{54},$$

regardless of the number of weeks since the last jackpot was awarded. ∎

The jackpot of over \$3.5 million was won following the drawing of January 8, 2023. The $A\spadesuit$ was revealed under card #49.

Table 2.14 collects some of the more useful probability and counting rules that we have discussed so far. In referring to this table, note that the First Addition Rule applies only to mutually exclusive events, while the Second Addition Rule applies to all events. Similarly, the first Multiplication Rule in the table requires that the events A and B be independent; the General Multiplication Rule applies in all cases.

TABLE 2.14: Common probability formulas

Complement Rule	$P(A^C) = 1 - P(A)$.
Subset Rule	If $B \subset A$, then $P(B) \leqslant P(A)$.
Permutations	$_nP_r = \dfrac{n!}{(n-r)!}$.
Combinations	$\dbinom{n}{r} = \dfrac{n!}{r! \cdot (n-r)!}$.
First Addition Rule (mutually exclusive events)	$P(A \text{ or } B) = P(A) + P(B)$.
Second Addition Rule	$P(A \text{ or } B) = P(A) + P(B) - P(A \text{ and } B)$.
Multiplication Rule (independent events)	$P(A \text{ and } B) = P(A) \cdot P(B)$.
Conditional Probability	$P(B|A) = \dfrac{P(A \text{ and } B)}{P(A)}$.
General Multiplication Rule	$P(A \text{ and } B) = P(A) \cdot P(B|A)$.

2.5   Exercises

Answers to starred exercises begin on page 285.

2.1. Use the Multiplication Rule to confirm that the probability of any one number being chosen in a round of card-based roulette at the Barona Casino (page 21) is still $\dfrac{1}{37}$ despite the curious setup. Note that the wheel spin and card draw are independent events.

2.2. Some local variations of poker recognize a hand called a *blaze*, which consists of five face cards. (Note that only jacks, queens, and kings are considered face cards.) A blaze beats two pairs but loses to three of a kind, although a blaze containing three of a kind or higher (for example, the full house $JJJQQ$) need not be called a blaze [33].

a. Find the probability of being dealt five face cards.

b. Find the probability that a player is dealt a hand that he or she calls a blaze—which must consist of two pairs of face cards and a fifth face card of the third rank, such as $QQJJK$.

2.3.[*] In [7], John Blackbridge asserts that a hand containing 5 cards of the same color should rank between 1 pair and 2 pairs. Find the probability that a five-card poker hand contains five cards of the same color.

We note here that some high-ranking poker hands—flushes and straight flushes, for example—are necessarily one-color hands that rank higher than 2 pairs and would be played as their standard value instead of as a one-color hand. This need not be considered here.

2.4.* Calculate the probability of the poker hand three of a kind. Note that in a hand holding three of a kind, if the other two cards form a pair, the hand is a full house; these need to be excluded from your calculations.

2.5. Consider a modified deck of cards containing six suits, with the usual 13 cards in each suit for a total of 78 cards. The two new suits are red *crowns* and black *anchors*.

a. In a five-card poker game played with this deck, five-of-a-kind would be a possible hand. Compute the probability of five-of-a-kind. Would five-of-a-kind outrank a royal flush?

b. Find the probability of being dealt a hand containing five cards of different suits, which is called a *rainbow*.

c. A *rainbow straight* is a hand consisting of five cards in sequence and all of different suits. Find the probability of a rainbow straight.

2.6. *MyDaY* is a Nebraska Lottery offering where players choose a date from the calendar instead of a number [9]. The state draws a single date in a daily drawing, and prizes are paid according to how many components of the date are matched. The date is composed of three separate choices:

- The month, which is a number between 1 and 12.

- The day, which is a number between 1 and 31. Only valid calendar dates are accepted in MyDaY, so the computer system will block attempts to wager on nonexistent dates such as November 31. February 29 may be chosen only if the year is a bona fide leap year, including the year 2000. In the history of MyDaY, beginning in October 2008, February 29 was drawn 4 times: twice in 2016 and twice in 2020. Coincidentally, both 2016 and 2020 were leap years.

- The year, which is a two-digit number from 00 through 99. These digits represent the last two digits of the year, with the understanding that 00 represents 2000 (which was a leap year, permitting bets on 2/29/00) rather than 1900 (which was not a leap year) [82].

a. How many dates are possible in MyDaY?

b. Matching the month alone wins a prize of $1–a break-even proposition for these $1 tickets. If you purchase a ticket for May 17, 1965, find the probability that you will match only the month.

c. A ticket that matches the month and day, but not the year, wins $12. For the ticket in part b., find the probability of winning the $12 prize.

2.7.* *2by2* is a game offered by the Multi-State Lottery Association, the organization responsible for Powerball, in Kansas, Nebraska, North Dakota, and Wyoming. 2by2 is effectively a double 2/26 lottery: players pick two numbers, in the range 1 to 26, in each of two colors: red and white. Two numbers in each color are drawn independently, and a ticket wins if at least one of its four numbers matches them. The payoff table for 2by2 is shown in Table 2.15.

TABLE 2.15: 2by2 payoff table [1]

Red matches	White matches	Prize
2	2	$22,000
2	1	$100
1	2	$100
2	0	$3
0	2	$3
1	1	$3
1	0	Free ticket
0	1	Free ticket

Due to symmetry, there are five different winning probabilities. Find each one.

2.8.* Repeat Example 2.1.14 using the 2004 Powerball configuration. Which event has a higher probability: matching 4 numbers without the Powerball or matching 3 with the Powerball?

2.9. The most recent U.S. state to establish a state lottery is Mississippi, whose state legislature approved lottery legislation in 2018. Mississippi Match 5 is a 5/35 game with three weekly drawings. Find the probability of winning a prize in this game by matching 2 or more numbers.

2.10.* The Ring of Fire keno bet (Example 2.1.19) offers a payout of $100,000 if 20 of the 32 edge numbers are chosen by the casino. Find the probability of this payoff.

2.11. Calculate the number of betting combinations on a keno way ticket consisting of two 2-spots, four 3-spots, and one 5-spot, with the understanding that no combination may involve more than 15 numbers.

2.12.* The Kewadin Casino in Sault Sainte Marie, Michigan offers a way bet called the "Cover All" keno wager. As the name suggests, the Cover All bet has all 80 numbers working for the player. In one Cover All game, the numbers are divided into 2×2 blocks of 4, according to how they are arranged on the bet slip (see Figure 2.3), so 1-2-11-12 is one block, as is 3-4-13-14, and so on up to 69-70-79-80. These 20 blocks are each combined with every other block to form virtual 8-spot tickets. Show that there are 190 distinct 8-spot combinations available in a Cover All game.

2.13. The "Top and Bottom" keno game at the Kewadin Casino does not require the player to choose any numbers. By playing a $5 Top and Bottom ticket, you win if 13 or more of the 20 numbers drawn fall on the top half of the ticket, which bears numbers from 1 to 40, or on the bottom half, where the numbers run from 41 to 80. Prizes are awarded according to how many numbers hit: a 13-7 or 7-13 top/bottom split pays off with a free ticket for another game, while a 20-0 or 0-20 split pays $25,000.

Find the probability of winning on a Top and Bottom ticket.

2.14. Recall the FireKeepers Casino's 20-spot $2 keno wager, whose top prize of $100,000 is won when the player matches 15 or more of the 20 drawn numbers.

a. Find the probability of hitting exactly 15 of 20 numbers in keno.

b. By mimicking the calculation in part a, find the probability of matching 16, 17, 18, 19, and 20 numbers.

c. What is the probability that a single keno ticket will win the $100,000 prize?

2.15. Suppose that on a football (soccer) weekend in Great Britain, 11 of the 49 games end in score draws. What is the probability that your ticket for ten games includes at least one line with eight score draws?

2.16. In December 2022, the sports books operated by MGM Resorts International offered the following futures odds on the College Football Playoff.

- Georgia: −165.
- Ohio State: 7–2.
- Michigan: 3–1.
- Texas Christian: 16–1.

Compute the over-round for this set of odds.

2.17. Video poker (see Section 3.2) is an electronic version of five-card draw poker. When the option to discard some cards and draw replacements is included, the probability of a royal flush is roughly 1 in 40,000. Use this estimate to calculate the probability of *not* drawing a royal flush in 40,000 video poker hands.

2.18. In Texas hold'em, suppose that your hole cards are an ace and a 2. Find the probability that the player to your immediate left has been dealt pocket aces.

2.19. If two pairs appear on the board in a hand of Texas hold'em, the board is said to be *double-paired*. This tends to limit the subsequent betting, as every player has at least 2 pairs and there's not a lot of room for improvement. Find the probability that a hold'em board is double-paired in the first 4 cards.

2.20.* Consider the following 2 events in a hand of Texas hold'em:

- A = Flopping a set (holding three-of-a-kind after the flop) given that you hold a pair.

- B = Flopping exactly 4 cards to a flush, given that your hole cards are suited.

Which event has the greater probability, A or B?

2.21. Repeat Exercise 2.20 allowing for the possibility of flopping 3 or four-of-a-kind in event A and flopping 4 or 5 cards to a flush in B, given the same hole cards.

2.22.* Two bingo cards with different sets of 24 numbers are called *disjoint*; cards with one or more numbers in common are said to *overlap*. Given the layout of a card, it is possible to construct three disjoint cards—the 72 numbers on the three cards include all 75 numbers except three of the numbers from 31 to 45, due to the free space under the N. In a Cover All game with these three cards, find the probability that there will be a winner by the time that the 65th number is drawn.

2.23. Find the probability that two randomly selected bingo cards will have no numbers in common.

2.24.* Bingo cards are typically sold in packages of 3000. Use the probability calculated in Exercise 2.23 to find the probability that a package will contain two or more disjoint cards.

2.25.* In the 2005 episode "Can You See What I See?" of the TV series *Las Vegas*, pit boss Nessa Holt states that she has seen a girl hit 56 straight passes [at craps], a player hit 17 straight naturals in blackjack, moving among 9 different tables, and the number 14 come up 14 straight times at roulette. Using the approximation $P(\text{Natural}) = \dfrac{1}{21}$ and assuming an American wheel, rank these 3 events by their probability from lowest to highest.

2.26. If you make the Ring of Fire keno bet (Example 2.1.19), you lose your wager if exactly 6, 7, 8, or 9 edge numbers are drawn. Find the probability of this event.

2.27.* Find the probability that each number on an American roulette wheel will come up exactly once in a string of 38 consecutive spins.

Chapter 3

Probability Distributions and Expectation

3.1 Random Variables

Definition 3.1.1. A *random variable* (RV for short) is an unknown quantity X whose value is determined by a chance process.

This is another definition that, on its face, isn't terribly useful—indeed, this comes perilously close to using the words "random" and "variable" in its own definition. Once again, a sequence of examples will illustrate this important idea far better than a formal definition.

Example 3.1.1. Roll 2d6 and let X denote their sum. X then takes on a value in the set $\{2, 3, 4, 5, 6, 7, 8, 9, 10, 11, 12\}$. ∎

Example 3.1.2. In a five-card poker hand, let X count the number of aces it contains. X can equal 0, 1, 2, 3, or 4. ∎

Example 3.1.3. In a hand of blackjack, let X denote the sum of the first two cards, counting the first ace as 11. Here, X is an integer between 3 and 21. (A hand containing two aces would be counted here as 12, not 2 or 22.) ∎

Example 3.1.4. Let X be the number spun on a European roulette wheel. Then X is contained in $\{0, 1, 2, 3, \ldots, 35, 36\}$, a set with 37 elements. ∎

Our next definition connects the possible values of a random variable with the respective probabilities of those values.

Definition 3.1.2. A *probability distribution* for a random variable X is a list of the possible values of X, together with their associated probabilities.

Example 3.1.5. Suppose we roll 2d6 and let X denote the sum of the numbers rolled. The possible sums are illustrated in Table 3.1, and the corresponding probability distribution for X is compiled in Table 3.2. ∎

Axioms 2 and 3 require that the sum of the probabilities in Table 3.2, and in any probability distribution, is 1.

DOI: 10.1201/9781003358183-3

TABLE 3.1: Sample space of outcomes when rolling 2d6

X	⚀	⚁	⚂	⚃	⚄	⚅
⚀	2	3	4	5	6	7
⚁	3	4	5	6	7	8
⚂	4	5	6	7	8	9
⚃	5	6	7	8	9	10
⚄	6	7	8	9	10	11
⚅	7	8	9	10	11	12

TABLE 3.2: Probability distribution when rolling 2d6

x	2	3	4	5	6	7	8	9	10	11	12
$P(X = x)$	$1/36$	$2/36$	$3/36$	$4/36$	$5/36$	$6/36$	$5/36$	$4/36$	$3/36$	$2/36$	$1/36$

To ensure that their craps tables are profitable, casinos depend on their dice following this probability distribution. In order to guarantee that the real dice conform to these theoretical probabilities, casino dice are manufactured to exacting specifications and are perfect cubes to within 1/10,000 of an inch on each side. The spots on the dice are made of a different-colored plastic with the same density as that used for the dice and are flush with the surface rather then being indented, as is the case with common game dice. Moreover, casino dice have razor-sharp edges and corners, in order to ensure that they bounce and roll in a truly random fashion. Common dice, such as those used for playing board games like backgammon, have rounded courses and edges so that they will roll better in the limited space for rolling dice. Figure 3.1 compares casino dice with board game dice.

FIGURE 3.1: Left: Casino dice from the Palms Casino in Las Vegas. Right: Common board game dice.

Additionally, casino dice are typically used on the tables for eight hours or less and are then removed from play before they develop small chips or other imperfections that might cause them to deviate from this distribution. Dice so removed are immediately canceled with a circle punched into one face or a hole drilled through two faces to guard against reintroduction of modified dice to the tables, and replaced by new dice. (In New Jersey, dice removed from play must be canceled by drilling.) Depending on casino policy, canceled dice may either be destroyed, given away to players, or offered for sale in the casino's gift shop.

A similarly short lifetime on the casino floor awaits playing cards, which are typically used for one eight-hour shift or less and then replaced before they have a chance to accumulate telltale nicks or marks that might tip off an alert gambler to the imminent arrival of a particularly high or low card. Cards used in baccarat or blackjack are also prone to develop "waves" from being bent back by players peering at the card faces, and these curves can also indicate a rank if a player is paying very close attention. Since this would lead to a departure from the true randomness on which the rules of the games depend, and since playing cards are relatively inexpensive, this is a small investment in ensuring that the games are being played correctly. Large casinos go through thousands of decks every week; the Bellagio in Las Vegas has a standing order with its card manufacturer for 60 cases of cards (8640 decks) per week [68]. As with dice, retired card decks are canceled, either by drilling (see Figure 3.2) or clipping off one corner, before being discarded or sold as souvenirs.

FIGURE 3.2: Used playing cards from FireKeepers Casino, Battle Creek, MI. The cards have been canceled with a hole drilled through their centers.

Example 3.1.6. *Door pops* are a relatively unsophisticated type of loaded dice. One die of the pair has a 2 on three sides and a 6 on the other three, and the other has a 5 on all six sides [103]. It follows that a pair of door pops can only roll a 7 or 11—and thus will always produce a winner on a come-out roll. It stands to reason that such a pair of dice, if introduced into a craps game, would be very easily discovered. On the other hand, the author has, on more than one occasion, presented a pair of door pops to a class of students and watched some of them closely examine the dice for several minutes before figuring out how they differ from standard dice.

If we define X to be the sum of the numbers showing when a pair of door pops is rolled, we have Table 3.3, the probability distribution for X.

TABLE 3.3: PDF for the sum when a pair of door pops is thrown

x	7	11
$P(X = x)$	$1/2$	$1/2$

■

Example 3.1.7. *Chuck-a-luck* is a game played with three dice. In its original form, it is not often seen in casinos anymore but can still be found in carnivals and, in expanded form, in the casino game *sic bo* (page 183). The three dice are tumbled in a wire cage. Gamblers bet on the numbers from 1 to 6, and are paid according to how many of their number appear on the dice: the amount wagered is matched for each die showing the selected number, so the payoff is 1 to 1 if one die shows the number, 2 to 1 if two do, and 3 to 1 if all three do.

Let X be the number of 4s that appear when the cage is spun. Since there are six sides on each die, there are $6 \cdot 6 \cdot 6 = 216$ ways that the dice can land. We note that the numbers showing on the three dice are independent random variables. We shall consider each possible value for X individually.

$X = 0$: There are five sides on each die that do not show a 4, so there are $5 \cdot 5 \cdot 5 = 125$ ways not to roll any 4s. Accordingly, $P(X = 0) = \dfrac{125}{216}$.

$X = 1$: For convenience, we shall imagine that the three dice are different colors: red, green, and blue. (In practice, the cage contains three identical dice.) Since there are five ways on each die not to roll a 4, and one way to roll a 4, we have the following chart of possibilities:

Red	Green	Blue	Count
4	Not 4	Not 4	$1 \cdot 5 \cdot 5 = 25$
Not 4	4	Not 4	$5 \cdot 1 \cdot 5 = 25$
Not 4	Not 4	4	$5 \cdot 5 \cdot 1 = 25$

Adding the last column shows that there are 75 ways to roll exactly one 4, so we conclude that $P(X = 1) = \dfrac{75}{216}$.

$X = 2$: Once again, we organize our work in a chart:

Red	Green	Blue	Count
4	4	Not 4	$1 \cdot 1 \cdot 5 = 5$
4	Not 4	4	$1 \cdot 5 \cdot 1 = 5$
Not 4	4	4	$5 \cdot 1 \cdot 1 = 5$

Adding the last column shows that there are 15 ways to roll exactly two 4s, so we have $P(X = 2) = \dfrac{15}{216}$.

X = 3: There is only one 4 on each die, so there is only one way to roll three 4s. We conclude that $P(X = 3) = \dfrac{1}{216}$. This probability could also have been calculated by using the complement rule and the three probabilities previously computed.

Combining this information gives us the probability distribution for X: Table 3.4.

TABLE 3.4: PDF for a spin of a chuck-a-luck cage

x	0	1	2	3
$P(X = x)$	$125/216$	$75/216$	$15/216$	$1/216$

■

Example 3.1.8. At FireKeepers Casino, a 20-spot keno game offers 19 ways to win. For a minimum bet of \$2, the player picks 20 numbers in the range 1–80 and wins unless 4 or 5 of the numbers selected are matched by the casino's numbers, although matching 3 or 6 numbers pays off only \$1—a net loss of \$1—and matching 2, 7, or 8 numbers merely gets the player their \$2 wager back. Let X count the number of matches. Since we account for every number drawn by the casino in computing keno probabilities, we have the following formula for $P(X = x)$, which could be used to compute the numerical probabilities that comprise a distribution:

$$P(X = x) = \frac{\dbinom{20}{x} \cdot \dbinom{60}{20 - x}}{\dbinom{80}{20}}.$$

In the numerator, the factor $\dbinom{20}{x}$ corresponds to the number of ways to choose x numbers from the 20 selected by the player. The second factor, $\dbinom{60}{20 - x}$, counts the number of ways to choose the remaining $20 - x$ numbers from among the 60 that the player did not choose. The denominator remains $\dbinom{80}{20}$, the number of possible subsets of 20 keno balls from a set of 80. ■

What can we conclude from this probability function?

- The probability of losing money by matching 3–6 numbers is

$$P(X = 3) + P(X = 4) + P(X = 5) + P(X = 6) = .7534,$$

which conforms to our notion that keno is not a very player-friendly game.

- Including the three break-even payoffs, the probability of not profiting from this bet is .9662, which definitely makes this look more like a casino game.

3.2   Expected Value and House Advantage

The notion of *expected value* is fundamental to any discussion of random variables and is especially important when those random variables arise from a gambling game. The expected value of a random variable X is, in some sense, an "average" value, or what we might expect in the long run if we were to sample many values of X.

The common notion of "average" corresponds to what mathematicians call the *mean* of a data set: add up all of the numbers and divide by how many numbers there are. For a random variable X, this approach requires some fine-tuning, as there is no guarantee that a small sample of values of X will be representative of the range of possible values. Our interpretation of average will incorporate each possible value of X together with its probability, computing what is in some sense a long-term average over a very large hypothetical sample.

Definition 3.2.1. The *expected value* or *expectation* $E(X)$ of a random variable X is computed by multiplying each possible value for X by its corresponding probability and then adding the resulting products:

$$E(X) = \sum_x x \cdot P(X = x).$$

This expression may be interpreted as a standard mathematical mean drawn from an infinitely large random sample. If we were to draw such a sample, we would expect that the *proportion* of sample elements with the value x would be $P(X = x)$; adding up over all values of x gives this formula for $E(X)$.

We may abbreviate $E(X)$ to E when the random variable is clearly understood. The notation $\mu = E(X)$, where μ is the Greek letter mu, is also common, particularly when the expected value appears as a term in another expression.

Example 3.2.1. Let X be the result when a fair d6 is rolled. The expectation is

$$E(X) = \sum_{x=1}^{6} x \cdot P(X = x) = \sum_{x=1}^{6} \frac{x}{6} = \frac{21}{6} = 3.5.$$

∎

The notation used in Definition 3.2.1 does not indicate the limits of the indexing variable x, as we have done with summation in earlier chapters; this is because those values may not be a simple list running from 1 to some n. When written this way, we should take this sum over *all* possible values of the random variable X, as in the next example.

Example 3.2.2. If we let X be the outcome when $1 is wagered on an American roulette corner bet (four numbers), the expression for $E(X)$ would be

$$E(X) = \sum_x x \cdot P(X = x),$$

where $x = -1$ or $x = 8$. The resulting expectation would be

$$E(X) = (8) \cdot \frac{4}{38} + (-1) \cdot \frac{34}{38} = -\frac{2}{38}.$$

∎

Definition 3.2.2. If X is a random variable measuring the payoffs from a game, we say that the game is *fair* if $E(X) = 0$.

Example 3.2.3. Suppose you gamble with a friend on the toss of a coin. If heads is tossed, you win $1; if tails is tossed, you pay $1. Since a fair coin can be expected to land heads and tails equally often, the expected value for this game is

$$E = (1) \cdot \frac{1}{2} + (-1) \cdot \frac{1}{2} = 0.$$

The game is fair. ∎

If a game is fair, then in the long run, we expect to win exactly as much money as we lose, and thus, aside from any possible entertainment derived from playing, we expect no gain. This is often summarized in the following principle:

If a game is fair, don't bother to play.
If a game is unfair, make sure it's unfair in your favor.

Failure to heed this maxim, of course, is responsible for the ongoing success of the gambling industry, for, as we shall see, games which are unfair and favor the gambler are rare.

Example 3.2.4. In 1998, Baldini's Casino in Sparks, Nevada ran a promotion aimed at local residents. It is common practice for casinos catering to local residents rather than tourists to offer paycheck cashing services, and anyone cashing their paycheck at Baldini's received a free spin on a special video poker machine. This machine offered a million-dollar payout to any player dealt a *sequential* royal flush—the ace, king, queen, jack, and 10 of the same suit, with all five cards in either ascending or descending order [78]. What does the million-dollar top prize contribute to the expected value of this game?

There is no cost to play this game, so this is one of those rare cases where the advantage lies with the player. However, having the advantage doesn't mean that it's in any way lucrative. Receiving a dealt royal flush is difficult enough; $p = \dfrac{4}{2,598,960}$. Getting the cards dealt in order is considerably more difficult.

The number of possible arrangements of 5 cards out of 52, where the order matters, is

$$_{52}P_5 = 311,875,200.$$

Of these hundreds of millions of arrangements, only 8 are sequential royal flushes: there are 4 suits to choose from, and 2 orders in which the royal flush can be arranged in sequence. The probability of a dealt sequential royal flush, therefore, is

$$\frac{8}{311,875,200} = \frac{1}{38,984,400}.$$

The last denominator is close to the 2020 population of California, making the probability of winning this game about the same as randomly choosing a resident of California and getting the attorney general.

We find that

$$E = (1,000,000) \cdot \frac{8}{311,875,200} = \frac{8,000,000}{311,875,200} = \frac{2500}{97,461} \approx \$.0257,$$

or approximately 2½¢. ∎

This was a free promotion, so there was no risk to players, but there was also very little risk to the casino offering this game. Other prizes were offered for different poker hands; the easiest one to win was a free beer [78]. Of course, if you offer paycheck-cashing services, then you have people in your casino with money in their pockets and a wealth of places where they can spend it, which explains what the casino has to gain from promotions like this one.

If someone were to offer you 3¢ for the opportunity to take your spin on the machine—ignoring all other payoffs—you would be wise to accept it, for they're offering you more than this chance is worth.

Example 3.2.5. Example 1.5.1 described the carnival game where the barker attempts to guess the month of your birth. Noting right away that your chance of winning is greater than your chance of losing, how does the carnival profit from this game?

The expectation of this game rests on two factors: the cost to play and the value of the prize. Let us denote these two values by C and V. If you win the game, you win a prize valued at V, but must subtract the cost C of playing from its value. If you lose, of course, your loss is $\$C$. The expectation from this game is then

$$E = (V - C) \cdot \frac{7}{12} + (-C) \cdot \frac{5}{12} = \frac{7V}{12} - C,$$

and as long as this value is negative, the carnival has an advantage. This happens when

$$V < \frac{12C}{7},$$

and so if, for example, the player is charged \$3 for a chance at a prize, the prizes can be worth as much as \$5.14 each and the carnival will still make money in the long run. ∎

An important principle of expected value, one that will be very important in the analysis of gambling systems, is contained in the following theorem.

Theorem 3.2.1. *If $X_1, X_2, \ldots, X_n$ are random variables, then*

$$E(X_1 + X_2 + \cdots + X_n) = E(X_1) + E(X_2) + \cdots + E(X_n).$$

Another way to state this result is to say that *expectation is additive*. The expected value of a sum of random variables—for example, a collection of outcomes of a sequence of bets—is the sum of the expected values of the individual RVs.

In a casino game, the expected value of a random variable has an important interpretation as the *house advantage* (abbreviated HA), or the percentage of each bet that a casino expects to win.

Definition 3.2.3. The *house advantage (HA)* of a game with a wager of N and payoffs given by the random variable X is $-\dfrac{E(X)}{N}$.

For convenience, we will assume that most bets are for \$1. This makes converting house advantages to percentages easier. For bets of more or less than \$1, the expected value can be found by multiplying both sides of the equation by the amount wagered, and the HA can be found by dividing the calculated expectation by the size of the bet.

As the term suggests, a fair game has zero HA. If the expected value is negative, then the HA is positive and the game favors the casino; a game with a positive HA favors the gambler. Very few casino games—blackjack with card counting, video poker under certain payoff tables and with perfect strategy, live poker, and sports betting—ever have a favorable long-term edge for the player. It should be noted that these four games all have a skill component. No standard casino game where the results are completely due to chance has a positive player expectation under ordinary playing conditions—although casino errors can create a lucrative opportunity for players; see page 184 for one such incident. Many of the house edges computed in this and the next chapter are collected in the appendix on page 283.

Roulette

Example 3.2.6. Let X be the random variable measuring the outcome of a \$1 bet on the number 17 at American roulette. The probability distribution for X is

$$\begin{array}{c|cc} x & 35 & -1 \\ \hline P(X=x) & 1/38 & 37/38 \end{array}$$

and the expected value of X is

$$E(X) = (35) \cdot \frac{1}{38} + (-1) \cdot \frac{37}{38} = -\frac{2}{38} \approx -\$.0526 = -5.26¢.$$

The house advantage is 5.26%. ∎

When gambling, $E(X)$ represents a "typical" outcome. In a single spin of a roulette wheel, we will never lose exactly 5.26¢ on a $1 bet, but if we make a large number of bets on many successive spins, we will find that our average loss per spin will be very close to this value.

Example 3.2.7. If we consider the bet in Example 3.2.6 on a European roulette wheel, the only change that needs to be made is switching 37 for 38 in the denominators to reflect the absence of a 00 on a European wheel. The resulting expectation is

$$E(X) = (35) \cdot \frac{1}{37} + (-1) \cdot \frac{36}{37} = -\frac{1}{37} \approx -\$.0270,$$

for an HA of 2.70%, about half the advantage of American roulette. While both games favor the casino, the player has a better chance of winning if the 00 is absent. ∎

In European roulette, the HA on even-money bets (red/black, high/low, and odd/even) is slightly lower than indicated in Example 3.2.7 when the *en prison* rule is used. Under this rule, if a player makes an even-money bet and the wheel turns up a 0, the bet is not resolved, but rather placed "in prison," to await the next spin. If the bet wins on the next spin, the initial bet is returned to the player with no additional payoff; if not, the bet is collected. If the second spin is also a 0, the bet loses—*en prison* lasts for only one additional spin.

The house advantage for a $1 even-money bet with *en prison* in force decreases, to

$$(1) \cdot \frac{18}{37} + (0) \cdot \frac{1}{37} \cdot \frac{18}{37} + (-1) \cdot \left(\frac{18}{37} + \frac{1}{37} \cdot \frac{19}{37}\right) = -\$0.0139,$$

and so the HA has dropped by nearly 50% on even-money bets, to 1.39%.

One might reasonably wonder why a casino would offer the *en prison* option, which works to reduce its edge. One possibility, as explained in [70], is that *en prison* is offered on the even-money bets, which are the bets most often involved in gambling systems for roulette (see Section 6.1). By decreasing the risk on even-money bets, the casino is subtly encouraging system players. Since betting systems do not work, this encouragement allows a casino to make up in volume what it gives back in a smaller HA.

More generally, for any bet on n numbers on an American roulette wheel except the basket bet ($n = 5$), we have the probability distribution shown in Table 3.5.

TABLE 3.5: General PDF for a roulette bet on n numbers, $n \neq 5$

x	-1	$(36-n)/n$
$P(X = x)$	$\dfrac{38-n}{38}$	$\dfrac{n}{38}$

This PDF corresponds to an expectation of

$$E = \left(\frac{36-n}{n}\right) \cdot \frac{n}{38} + (-1) \cdot \frac{38-n}{38} = -\frac{2}{38}.$$

This is independent of n and shows that the casino has a constant 5.26% edge on every roulette bet other than the basket bet.

With this house advantage on every bet except the basket bet, where the HA is 7.89%, American roulette is one of the worst games in a casino for players. It is an excellent example of the gambler's rule of thumb expressed on the "Luck Be An Old Lady" episode of *Sex and the City* (2002):

> *The easier a bet is to understand, the greater the house edge.*

It's not difficult at all to understand the idea of "pick some numbers and see if they come up on a wheel." By contrast, the house advantage on a blackjack hand, for a player using perfect basic strategy, is approximately .5% (depending on the number of decks used and the exact house rules), and the HA on the pass line bet in craps is only 1.41% (see Example 3.2.9). The calculations leading to these values are somewhat more complicated than those above, as are the rules of the games.

If the wheel has 0, 00, and 000 spaces, we can derive another formula similar to those above for the expectation of a wager on n numbers paying off at $\dfrac{36-n}{n}$ to 1. Rather than do so, we shall derive a general formula for a roulette wheel with z zeroes, for which the games we have so far considered emerge as special cases. Excepting the basket bet as usual, the expectation for a bet on n numbers is then

$$E = \left(\frac{36-n}{n}\right) \cdot \frac{n}{36+z} + (-1) \cdot \frac{36-n+z}{36+z} = -\frac{z}{36+z}.$$

Sands Roulette, with $z = 3$, offers the casino a consistent 7.69% edge on every bet except the basket bet. Adding a 0000 space to the wheel and the layout would round the HA up to an even 10%.

As bad as Sands Roulette might be for the gambler, other casino games can be far worse. The Match 1 bet at keno typically pays 3 for 1 and has an expected value of

$$E = (2) \cdot \frac{20}{80} + (-1) \cdot \frac{60}{80} = -\$0.25,$$

leading to a 25% house advantage. How many zeros would a roulette wheel have to have before the HA reached this level?

We seek to solve the equation

$$E(z) = \left(\frac{36-n}{n}\right) \cdot \frac{n}{36+z} + (-1) \cdot \frac{36-n+z}{36+z} = -\frac{z}{36+z} = -.25,$$

or

$$4z = 36 + z,$$

to which the solution is $z = 12$. A Match 1 keno ticket is equivalent to playing roulette with a 000000000000 pocket on the wheel.

The Big Six Wheel

Is the Big Six wheel (page 16) lucrative? Only for the casino, as we might expect given the simplicity of the game. The standard Las Vegas rules for the Big Six wheel state that each currency amount pays its face value per \$1 bet, and the two logos each pay off at 40 to 1 [113]. Armed with these payouts, we can see how much of an edge the casino has. For a \$1 bet on the \$1 spot:

$$E = (1) \cdot \frac{24}{54} + (-1) \cdot \frac{30}{54} = -\frac{6}{54} = -\frac{1}{9} \approx -\$.1111.$$

This 11.11% house advantage is the best bet, for the player, on the Big Six wheel. Other bets have HAs ranging from 16.67% to 24.07% (see Exercise 3.19).

In New Jersey, state regulations require that the two logos pay off at 45 to 1, rather than 40 to 1. While the HA on these bets drops from 24.07% to 11.11%, a 2011 study [107] showed that one casino's hold (the percentage of wagered money that is kept by the casino) at the Big Six wheel averaged 42.25%, well in excess of even the highest house advantage. This increased casino hold arises due to players re-betting their winnings and thus exposing more money to the already-high house advantage.

This study confirmed that Big Six is not a player-friendly game in practice, as well as in theory.

Dice Games

Example 3.2.8. In the game of chuck-a-luck described in Example 3.1.7, a \$1 bet pays \$1 for each time the selected number appears on the dice, so a player can win \$1, \$2, or \$3—or lose \$1. A winning player's original bet is returned, so we can modify the probability distribution for X, the number of times the selected number appears, to a distribution (Table 3.6) for the new random variable Y that counts a player's winnings. If $X = 0$, then $Y = -1$; all other values for Y are identical to those for X.

TABLE 3.6: PDF for a single-number chuck-a-luck bet

y	-1	1	2	3
$P(Y = y)$	$125/216$	$75/216$	$15/216$	$1/216$

The expected value of a \$1 bet on a single number in chuck-a-luck is thus

$$E(Y) = (-1) \cdot \frac{125}{216} + (1) \cdot \frac{75}{216} + (2) \cdot \frac{15}{216} + (3) \cdot \frac{1}{216}$$
$$= -\frac{17}{216} \approx -\$0.0787,$$

indicating a house advantage of about 7.87%. ∎

This may be the reason for chuck-a-luck's disappearance from casinos—if the house advantage is too high, players will eventually avoid the game. Individual bets with a high house advantage are not a barrier to long-term player acceptance of a casino game—many bets on the craps table, for example, have edges higher than 7.87%. In chuck-a-luck, however, the only bets available are single-number bets, and so all bets have the same 7.87% house edge. Without a less unfavorable betting option, chuck-a-luck has little to offer a perceptive gambler.

Example 3.2.9. In craps, we saw in Example 1.5.2 that the probability of winning a bet on the pass line is $p = \frac{244}{495} \approx .4929$. For a \$1 bet, the expected return is

$$E = (1) \cdot \frac{244}{495} + (-1) \cdot \frac{251}{495} = -\frac{7}{495} \approx -\$.0141 = -1.41¢,$$

so the house advantage is a mere 1.41%. The corresponding expectation for the don't pass bet, incorporating the Bar 12 rule, is

$$E = (1) \cdot \frac{949}{1980} \cdot + (0) \cdot \frac{1}{36} + (-1) \cdot \frac{244}{495} = -\frac{3}{220} \approx -\$.0134 = -1.34¢,$$

so the HA is 1.34%. ∎

These results show that, from a purely mathematical standpoint, the house advantages on pass and don't pass bets are very low, and these bets stand among the best in the casino for the gambler. These low house advantages can be reduced further by the use of the *free odds* betting option, which we shall consider in Chapter 4.

Craps: One-Roll Bets

Many of the other bets on a standard craps layout are simple one-roll bets that are resolved on the next roll of the dice. These bets have the advantage of being easy to understand, but as we might now expect, their house advantage is large.

Example 3.2.10. The "Any Seven" or "Big Red" bet is a one-roll wager that the next roll of the dice will show a 7. This bet pays off at 4 to 1 (or 5 for 1, as shown in Figure 1.5), and the probability of winning is just the probability of rolling a 7: $\frac{1}{6}$. The expected return on a $1 Any Seven bet is

$$E = (4) \cdot \frac{1}{6} + (-1) \cdot \frac{5}{6} = -\frac{1}{6} \approx -\$.1667,$$

so the HA is 16.67%.   ∎

The alternate name "Big Red" is a pretty good description of where your bankroll is headed if you make this bet, and in light of the other available bets with less onerous HAs, this one should be avoided.

Example 3.2.11. In Example 1.5.4, we looked at the C&E bet, which combines the Any Craps and Eleven bets. Assuming a $2 bet split evenly between the two wagers, what is the HA of C&E?

The Any Craps bet pays off at 8 for 1 (or 7 to 1) while Eleven's payoff is 15 for 1 (14 to 1). Note that if one of the two bets in this combination wins, the other loses; this must be accounted for in the calculations.

If a 2, 3, or 12 is rolled, the net win is $7 − $1, or $6. This event has probability $\frac{4}{36}$. If the dice show 11, the net payoff is $14 − 1 = $13, with probability $\frac{2}{36}$. The probability distribution for the net winnings X on the C&E bet is shown in Table 3.7.

TABLE 3.7: PDF for the craps C&E bet

x	6	13	−2
$P(X = x)$	$4/36$	$2/36$	$30/36$

The expectation on a C&E bet is then

$$E = (6) \cdot \frac{4}{36} + (13) \cdot \frac{2}{36} + (-2) \cdot \frac{30}{36} = -\frac{10}{36} \approx -\$.2778.$$

Dividing this expectation by the $2 wagered gives an HA of approximately 13.89%—better than Any Seven, but still too high for the serious craps player.   ∎

Example 3.2.12. Another one-roll bet popular on craps tables is the *field* bet. One version of the Field bet section of the layout is shown in Figure 3.3.

FIGURE 3.3: Field betting area [60].

While the specific details may vary slightly among casinos, each Field bet assembles a collection of less-likely outcomes and pays off if any one of the numbers is rolled. In Figure 3.3, the Field bet covers 2, 3, 4, 9, 10, 11, and 12. A casual observer might note that this includes 7 of the 11 possible sums, not noticing right away that these 7 numbers combine for a cumulative probability of only $\frac{16}{36}$, less than 50%. Some tables exchange the 9 shown in Figure 3.3 for the 5, which does not change the probability of winning.

On the surface, this looks like a good bet—a field bet covers 7 of the 11 possible sums and pays double or triple on two of them. Of course, as we know, this apparent advantage is based on an incorrect interpretation of the sample space (see Example 1.4.2). On a $1 bet:

- There is one way to win $3: by rolling 6-6.

- There is one way to win $2: by rolling 1-1.

- There are 14 ways to win $1: when the roll is a 3 (two ways), 4 (three ways), 9 (four ways), 10 (three ways), or 11 (two ways).

- The remaining 20 rolls—totals of 5, 6, 7, or 8—result in a $1 loss.

With the facts before us, we can derive the probability distribution for X, the net amount won on a $1 field bet:

x	-1	1	2	3
$P(X = x)$	$20/36$	$14/36$	$1/36$	$1/36$

and compute the expectation:

$$E = (3) \cdot \frac{1}{36} + (2) \cdot \frac{1}{36} + (1) \cdot \frac{14}{36} + (-1) \cdot \frac{20}{36} = -\frac{1}{36} \approx -\$.0278.$$

The HA on this bet is therefore 2.78%—exactly one-sixth the HA of "Any Seven." ∎

Alternate versions of the Field bet are available to casinos; these usually involve varying the bonus payout on 2 or 12 but do not change the overall probability of winning. The bonus payout affects the house edge.

- A 2–1 payoff on either 2 or 12 with an even-money payoff on the other raises the HA to 8.33%; offering this payoff on both 2 and 12 gives a 5.56% house advantage.

- The same 5.56% HA occurs if one of 2 and 12 pays even money and the other pays 3–1.

- If a casino offers 3–1 on both 2 and 12, the bet is fair, with no advantage on either side. This is unlikely in casino play unless someone has made an error in designing or printing the layout. [10].

A simple 1–1 payoff on all Field numbers was in force at a casino in the Bahamas in the 1970s with no bonuses on either 2 or 12 [51]. This version of the Field bet wins even money on 16 rolls and loses on the other 20, raising the casino's advantage to 11.11%—double the HA of a Field bet paying 2–1 on both 2 and 12.

Craps: Multi-Roll Bets

A second class of craps bets can extend over several rolls before being resolved. The *hardway* bets are one example. A hardway bet can be made at any time, whether or not a point has been established. A player making a hardway bet on an even number (4, 6, 8, or 10) bets that that number will be rolled "the hard way," as doubles, before it is rolled another way or a 7 is rolled.

A hardway bet can be won if any of the following sequences of events happens:

- The number is rolled as doubles on the first roll after the bet is made.

- Neither the number nor a 7 is rolled on the first roll, and the number is rolled as doubles on the second roll.

- Neither the number nor a 7 is rolled on the first two rolls, and the number is rolled as doubles on the third roll.

- Neither the number nor a 7 is rolled on the first three rolls, and the number is rolled as doubles on the fourth roll.

- And so on, through an infinite number of possibilities. As the number of rolls increases, of course, the probability of needing that many rolls to resolve the bet decreases.

Since the successive rolls are independent, we could use the Multiplication Rule to compute the probability of each of these sequences and the mathematical theory of infinite series, which is beyond the scope of this book, to add them all up. As we saw with the Pass and Don't Pass bets in Chapter 1, an easier approach to computing probabilities for hardway bets may be used. It focuses only on the rolls that will resolve the bet.

Example 3.2.13. Consider the case of a hardway bet on 6, which pays off at 9 to 1. We can ignore all possible rolls except 6s and 7s—there are 11 of these to be considered. Only one of them, the 3-3, results in this bet winning, so the probability of winning the bet is $\frac{1}{11}$. The expected value of a \$1 bet is then

$$E = (9) \cdot \frac{1}{11} + (-1) \cdot \frac{10}{11} = -\frac{1}{11} \approx -\$.0909,$$

for a 9.09% house advantage. This bet may be better than "Any Seven," but there are still far better bets available. Even the field bet is a better bet for the player. ∎

Since 6 and 8 are equally likely, the HA of a hardway bet on 8 is also 9.09%. Hardway bets on 4 or 10 pay off at 7 to 1, a lower payoff since there are fewer "easy" ways to roll these numbers. Confining our attention to the rolls that resolve the bet, we find that the probability of winning this bet is $\frac{1}{9}$, from which it follows that the house edge is 11.11%.

One of the most attractive craps bets to players is not listed on the layout. *Place bets* get their name from the fact that they are made for a player by a dice dealer; you place your chips in a neutral place on the layout and tell the dealer how you want them wagered. A place bet may be made on any point number and is simply a bet that that number will be rolled before a 7. This differs from a pass line bet in that the number you choose need only be rolled once, not once to establish the point and then a second time to win. Place bets pay off as follows:

- 4 and 10 pay off at 9 to 5.
- 5 and 9 pay 7 to 5.
- 6 and 8 pay 7 to 6.

As always, fractions are rounded down, in favor of the house, so the wise gambler will make place bets in multiples of 5 or 6, as appropriate.

The casino can afford to pay place bets at slightly better than even-money odds because a player making a place bet is sacrificing the chance to win a pass line bet on the come-out roll, and this gives up more than the player gains by not running the risk of losing on the come-out. Since 6s and 8s are equally likely when 2d6 are rolled, place bets on those numbers are mathematically equivalent. As with Pass and Don't Pass, we can restrict our attention to the rolls that resolve the bet: there are 5 ways to roll a 6 and 6 ways to roll a 7. The expected value of a \$6 bet is

$$E = (7) \cdot \frac{5}{11} + (-6) \cdot \frac{6}{11} = -\$\frac{1}{11} \approx -\$0.0909.$$

Dividing by the wager reveals that the HA of a place bet on 6 is only 1.52%.

The *lay bet*, also absent from the layout, is the equivalent of the place bet

for gamblers wishing to bet against the shooter. Lay bettors are wagering that
a 7 will be rolled before the number they choose appears. Since betting on the
7 gives the lay bettor an advantage, lay bets pay off at less than even money.
A lay bet on the numbers 4 or 10 pays out at 1–2, 5 or 9 pay 2–3, and 6 or 8
pay 5–6. These are the true odds, so the casino has no advantage yet. Casinos
charge a 5% commission on lay bets, so a player must pay an additional $1
for every $20 that he stands to win on a lay bet. This commission may be
charged on all lay bets, or only collected from winning lay bettors—which
makes a difference in the expectation and the HA.

Laying the 6, for example, is a wager that a 7 will be rolled before a 6. As
with placing a 6, there are only 11 rolls that matter to resolving the bet. If
the bet is made for $24, the player hopes to win $20, so the commission is $1
and the bet must be made for $25 if the commission is charged on all bets.
The $1 commission is lost regardless of the outcome of the bet, so the two
possible outcomes are "Win a net total of $19" and "Lose $25". The expected
value of this bet is

$$E = (19) \cdot \frac{6}{11} + (-25) \cdot \frac{5}{11} = -\$1$$

and the HA is 4.00%.

Lay bets charging a 5% commission should be made so that the amount to
be won is a multiple of $20, since casinos frequently round fractions in their
favor. If the lowest-value chip in the rack is $1, a $10 lay bet would be charged
$1, a 10% commission. For lay bets on 4 or 10, this means betting in multiples
of $40 before adding in the commission. Lay bets on 5 or 9 should be made
in multiples of $30, and lay bets on 6 or 8 should be multiples of $24. If the
casino uses 50¢ chips or half-dollar coins, these multiples may be halved.

Bingo, Keno, and Lotteries

In an actual Cover All bingo game (Example 2.4.12), it is highly unlikely
that a game will last until all 75 numbers are drawn. This would require that
every player in the game have a number common to all of their cards, and
that that number be the last one drawn—which would result in a massive tie
as everyone shouted "Bingo!" at once. We can compute the average number
of draws required for a given card to be covered by evaluating the sum

$$\sum_{k=24}^{75} k \cdot P(\text{Win in } k \text{ draws}) = \sum_{k=24}^{75} k \cdot \frac{24 \cdot 51! \cdot (k-1)!}{(k-24)! \cdot 75!} = 72.96.$$

If 200 cards are in play, then repeated simulation suggests that it will
take about 62 numbers before a winner is determined. This number slowly
decreases with the number of cards in a game; for 1000 cards, the average is
58.85 [119].

If a game does not refund the original bet to winning players, we must take this into account when computing expected returns. This is the case, for example, when playing keno or buying a lottery ticket. One way to do this is to reduce the advertised payoffs by the cost of the wager, and the next few examples demonstrate this approach.

Example 3.2.14. The Michigan State Lottery's Club Keno game, played in bars across the state, is an electronic version of keno. The 1 Spot Game asks the player to pick one number in the range from 1 to 80. The state then draws 20 numbers in that range, and if the player's number is among the 20, the payoff is $2, making the payoff odds 1 for 1. This is less than the standard 3 for 1 payoff typically offered in Match 1 games of live keno.

The probability of winning this bet is the probability of one's number appearing as one of the 20 drawn. This chance can be computed without the need for combinations by performing a thought experiment. Imagine that you buy 80 Club Keno tickets, one on each number. When Michigan draws its numbers, 20 of your tickets will be winners and 60 will be losers, so the probability that any one ticket is a winner is

$$\frac{20}{80} = \frac{1}{4}.$$

The probability of losing is therefore $\frac{3}{4}$. If tickets cost $1, which is paid when the ticket is purchased, then the appropriate probability distribution for your winnings X is

x	-1	1
$P(X = x)$	$3/4$	$1/4$

where each of the outcomes—$0 or $2—has been reduced by the cost of the ticket. The corresponding expected value is

$$E = (-1)\left(\frac{3}{4}\right) + (1)\left(\frac{1}{4}\right) = -\$.50 = -50\text{¢}$$

—giving a 50% HA which makes this bet equivalent to asking someone for change for a dollar and receiving two quarters in return. And being happy with the exchange. ∎

In 2022, the Wyoming State Lottery introduced a live keno game that paid $2.50 for a winning 1-spot ticket. This game has the slightly better, but by no means player-favorable, house advantage of 37.5%.

Example 3.2.15. For the Pick 7 keno game described in Example 2.1.15, we saw that the probability of winning was .0616. Of course, part of the lure of keno is the potential to win a large payoff for a small bet, as when matching seven out of seven numbers (an admittedly rare occurrence) pays off $10,000 for every dollar bet. Do the higher payoffs for the rarer events result in a reasonable expectation?

Let Y be the net amount won on a $1 bet, taking the cost of the ticket into account. The probability distribution for Y is

y	-1	0	19	349	9999
$P(Y = y)$	.9384	.0522	.0086	.00073	2.44×10^{-5}

The corresponding expectation is

$$E = (-1)(.9384) + (19)(.0086) + (349)(.00073) + (9999)(2.44 \times 10^{-5})$$
$$\approx -\$.2762,$$

so the house advantage is 27.62%, and while there is no well-defined meaning of "reasonable bet," it is safe to say that this bet does not qualify. ∎

Live keno is in decline in casinos, in part due to the fact that it's labor-intensive and requires a high HA to make up for the fact that live keno only draws about 12 games per hour. This high house edge can drive patrons to other games. At the same time, though, video keno, which can run through a game in well under a minute, is rising in popularity. The increased speed and the automatic nature of video keno can allow casinos to offer better odds with lower house advantages, confident that the money will continue to roll in through increased decisions per hour.

Table 3.8 compares live and electronic Pick 4 keno games at the FireKeepers Casino. While the payoff for catching 4 out of 4 numbers has been sharply

TABLE 3.8: FireKeepers Casino Pick 4 keno pay tables

Catches	Live payoff	Electronic payoff
2	1	2
3	2	5
4	160	91

reduced, this is the least-likely outcome and is more than compensated for by the larger payments for fewer catches. The HA for the live game is 21.07% and the edge for the electronic game is 7.97%: perhaps not a lucrative bet in its own right, but a lot lower than the corresponding live bet's house advantage. The machine with this game was in the same room where live casino draws are conducted, so a short walk would have uncovered a better game.

Another option when incorporating the cost of the wager into an expected value calculation is to work with the payoffs as advertised and then, at the end, to subtract the price of the ticket from the sum. The following two examples illustrate this method.

Example 3.2.16. A *50/50 drawing* is a common version of a lottery, often used as a charitable fundraiser, where 50% of the money raised by selling tickets is returned as a prize to the holder of the winning ticket. If the tickets are sold for \$1 each, and n tickets are sold, then the prize will be $\$\frac{n}{2}$ and the expected win when buying a single ticket is

$$E = \left(\frac{n}{2}\right) \cdot \frac{1}{n} = \$.50.$$

From this amount, which is independent of the number of tickets sold, we subtract the cost of the ticket to arrive at a final expectation of $-\$.50 = -50¢$. ∎

An expected value of half the ticket price is what we would expect, given the prize structure of a 50/50 drawing. This extends to the case where you purchase k tickets instead of just 1; while your probability of winning rises to $\frac{k}{n}$, you must subtract $\$k$ at the end rather than just \$1, and so your expectation is $-\$\frac{k}{2}$, or half of the money you paid out at the start.

Example 3.2.17. In the FireKeepers Casino's Pick 20 keno game, (Example 3.1.8), the expected *number* of matches is

$$\sum_{x=0}^{20} x \cdot P(X = x) = \sum_{x=0}^{20} x \cdot \frac{\binom{20}{x} \cdot \binom{60}{20-x}}{\binom{80}{20}} = 5,$$

and so it is no surprise that the casino has tagged this event, and three very near it, for a player loss. The expected value of a \$2 bet can be computed by using the payoff table for this game, Table 3.9.

TABLE 3.9: FireKeepers Casino: Pick 20 pay table, \$2 ticket

Matches	0	1	2	3	4	5	6
Payoff	400	4	2	1	0	0	1
Matches	7	8	9	10	11	12	13
Payoff	2	2	4	20	75	400	2000
Matches	14	15	16	17	18	19	20
Payoff	50,000	———	100,000	———			

Denoting this payoff function by $A(x)$, where x is the number of matches, and subtracting \$2 at the end to incorporate the cost of the ticket into the expectation, we find that

$$E = \left[\sum_{x=0}^{20} A(x) \cdot P(X = x)\right] - 2 \approx -\$0.3402,$$

which, upon dividing by the cost of the ticket, gives an HA of 17.01%. ∎

Example 3.2.18. The Michigan State Lottery also offers a "Daily 3" game, for which numbers are drawn twice a day, seven days a week. The simplest bet is a "straight" bet, where the gambler picks a three-digit number and wagers $1. If the player's number matches the three-digit number drawn by the state, the payoff is 500 for 1. Notice that the payoff is "500 *for* 1" rather than "500 *to* 1" and so the profit for a winning player is only $499. Since there are 1000 possible three-digit numbers, from 000 through 999, the probability of winning this lottery is $\dfrac{1}{1000} = .001$. The expected value of a $1 bet is

$$E = (499) \cdot \frac{1}{1000} + (-1) \cdot \frac{999}{1000} = -\frac{500}{1000} = -\$.50$$

or, subtracting the cost of the ticket at the end rather than reducing the payoffs,

$$E = (500) \cdot \frac{1}{1000} + (0) \cdot \frac{999}{1000} - 1 = -\frac{500}{1000} = -\$.50.$$

Either way, the state takes half of every dollar wagered on a Daily 3 ticket. ∎

Example 3.2.19. Another Daily 3 option is the box bet. As in horse racing (Example 2.1.8), when a player "boxes" a three-digit number, the ticket wins if the three digits comprising that number come up, in any order, in the state's chosen number. For example, if you make a box bet on the number 678, you win if the state's number is 678, 687, 768, 786, 867, or 876. The payoff structure depends on whether the player's number contains two or three different digits; we shall consider the case where the three digits are all distinct. This leads to what is called a "six-way boxed bet," since there are $_3P_3 = 6$ different permutations of a three-digit number consisting of three different digits. This bet pays off at 83 for 1 if any of the six permutations is drawn, and so the expected value is

$$E = (82) \cdot \frac{6}{1000} + (-1) \cdot \frac{994}{1000} = -\frac{502}{1000} = -\$.502,$$

slightly, though not significantly, less than the expectation for a $1 straight bet. ∎

While the lottery bets considered in Examples 3.2.14, 3.2.18, and 3.2.19 are exceptionally bad bets from the gambler's perspective, one fact that must be addressed here is that state lotteries are intended in part as a fundraising mechanism. Since its inception in 1972, the Michigan State Lottery has contributed over $26 billion to the state's public schools, and that number is far greater than it would be if players were getting a 95% return on their lottery dollars [81]. A 50% house advantage would be untenable in a casino, but is readily accepted in a state lottery when it is understood that a portion of the proceeds is supporting education.

Numbers and Policy Games

Daily state lottery drawings can trace their origins to a collection of underground games loosely connected under an umbrella called the *Numbers Game*. This name describes several versions of an illegal game of chance popular in large American cities during the middle of the twentieth century[104]. The Numbers Game is a very simple proposition: players bet on a number of their choice and win their bet if the number appearing in some legitimate and presumably unbiased source matches theirs. In its heyday, the Numbers Game attracted millions of dollars in bets and engaged a significant fraction of the adult population in some areas. While the game promoters kept very few reliable records, it was estimated that there were 500,000 daily bettors in New York City in the 1960s [43]. The game was popular in part because the cost of entry was tiny: in some locations, action could be had for as little as a penny.

Often, Numbers Games were run by local organized crime operations [43]. This made it necessary for transparency in determining the winning number, so that bettors could be assured that they were playing a game where they had at least a chance of winning. Some games generated winning numbers by using digits from the U.S. Treasury balance as published in a newspaper or a formula involving the payoff prices on specified races at an indicated racetrack; the latter method led to a winning number called the *Manhattan* number in New York City [71]. Also in New York, the *Brooklyn* number used the last 3 digits of the day's handle at Aqueduct racetrack [43, 61].

Example 3.2.20. Suppose that a round of the Numbers Game was played using the Manhattan number computed from the returns of the first 7 races in the Breeders' Cup on November 5, 2022. The payoffs from those races are shown in Table 3.10.

TABLE 3.10: Breeders' Cup results, November 15, 2022.

Race	1	2	3	4	5	6	7
Win	5.70	87.78	6.32	10.38	13.10	4.76	7.76
Place	4.76	20.76	4.14	4.84	19.64	33.28	12.40
Show	6.56	6.42	6.30	14.08	2.54	7.28	4.66
Total	17.02	114.96	16.76	29.30	35.28	45.32	24.82

Sums:	Races 1–3: 148.74
	Races 1–5: 213.32
	Races 1–7 : 283.46

The digits of the day's winning number, 833, are indicated in bold. The first digit is the units digit in the sum of the payoffs from races 1–3, the second is the units digit in the sum of the payoffs from the first 5 races, and the final digit is determined by adding the totals from races 1–7. ∎

A common form of the Numbers Game called on players to pick a 3-digit number. This permits 1000 possible choices. Several betting options were available, depending on the neighborhood [61].

- A *straight* bet typically paid 600 for 1 if the player's choice matched the chosen number. Some combinations—the most popular ones, such as 111, 222, or 711—were designated as restricted numbers, and payoffs were lower; as low as 450 for 1. This decreased payoff insulated the bank operating the game against financial ruin if a very popular number was selected as the winner.

- The *combination* bet was a box bet that paid 100 for 1 if the player's 3 digits appeared, in any order, in the winning number.

- Players could choose a *front bolita* or *back bolita* bet, which involved choosing a 2-digit number. Both bets paid winners at 60 for 1: the front bolita bet paid if the player's 2-digit number matched the first 2 digits of the winning number and back bolita paid off if the last 2 digits matched. Either bolita bet had expected value of

$$E = (60) \cdot \frac{10}{1000} - 1 = -0.40 \text{ units,}$$

and the game operators enjoyed a 40% advantage.

- *Single action* bets could be made on any one digit. Players specified which number they wanted and which position in the winning number they thought it would appear. These bets could pay 6, 7, or 8 for 1; with a probability of winning of $\frac{1}{10}$, the operators held an edge from 20–40%.

The house advantage on a straight bet seems like it should be 40%, provided that the winning number is not restricted, but the payoff mechanism decreased the payoff and raised the advantage. A player winning a 600 for 1 shot collected only 540 units; 10% of the win was taken as a tip by the collector, who served as the intermediary between gamblers and game operators. Collectors were responsible for collecting bets and dispensing winnings. The eventual expected value for a 1-unit bet was then

$$E = (540) \cdot \frac{1}{1000} - 1 = -0.46 \text{ units,}$$

and the HA was 46%, not 40%.

Despite the long odds, the Numbers Game was quite popular. In Detroit, an employee at the *Free Press* once clipped the two lines containing the day's Treasury balance from the business pages so that the stock market tables would all fit. When the newspaper hit the streets the next morning, the paper's telephone lines were swamped with calls wanting to know where the numbers were [27].

Daily numbers games were a target of state government-run state lotteries, which returned in the USA beginning in New Hampshire in 1963. The effect on numbers games was minimal early on, due to the fact that early state lotteries had no mechanism to allow players to choose their own numbers. These games, where tickets are sold in random or sequential number order, are known as *passive* lotteries. Over the course of a decade, lottery officials in several Eastern and Midwestern states sought ways to offer games with the features of the Numbers Game, with its daily drawings and player-chosen numbers. By 1977, multiple state lotteries had begun to offer daily draw games [141].

An alternate form of the Numbers Game was called a *policy game*. This game, popular in Chicago, Illinois, used a random drawing rather than a publicly available source to select winning numbers. Many of these games selected winning numbers in the range from 1–78; some operators used numbers drawn at a central location in Covington, Kentucky and transmitted by telegraph [71].

Betting options for policy games varied greatly by location and over time. A standard game setup called for gamblers to pick 3 numbers, giving them

$$\binom{78}{3} = 76,076$$

choices, while operators chose 12. This ticket, called a *gig*, paid 200 for 1 if the chosen numbers appeared anywhere on the list of 12. The probability of a winning ticket was then

$$\frac{\binom{75}{9}}{\binom{78}{12}} = \frac{5}{1729} \approx .0029.$$

A *flat gig* bet was a wager that the 3 chosen numbers appeared in 3 specified locations among the operator's dozen [71]. This more difficult feat carried a larger payoff of 1000 for 1. Without loss of generality, we can simplify our thinking by assuming that the three numbers are to appear in positions 1, 2, and 3, since the exact positions do not affect the probability of the 3 chosen numbers falling into them. Since the order in which the 3 numbers were drawn into the 3 positions did not matter, the chance of a winning flat gig ticket was

$$\frac{1}{\binom{78}{3}} = \frac{1}{76,076}.$$

which results in a huge house advantage, over 98%.

Another policy option was a *day number* bet, which called for the player to choose 1 number and paid 5 for 1 if it hit. As in Example 3.2.14, this bet can be analyzed without the need for advanced counting techniques. Assume that a bettor has made separate day number bets on each of the 78 numbers. When the numbers are drawn, 12 tickets will be winners, so the probability of winning on a single number is $\frac{12}{78} = \frac{2}{13}$. The expected value of a 1-unit ticket is

$$E = (5) \cdot \frac{2}{13} - 1 = -\frac{3}{13} = -.2308,$$

and the operator's edge was 23.08%.

The *Chicago Wheel* was a variation on the policy game. Players chose 3 numbers from 1–78, and game operators selected two lists, called columns, of 12 numbers each [19]. Drawings were held twice daily; to win, all 3 of the gambler's numbers had to appear among the 24 numbers drawn. For the moment, we are not concerned with the allocation of numbers to columns, so the probability of a winning ticket is then

$$p = \frac{\binom{3}{3} \cdot \binom{75}{21}}{\binom{78}{24}} = \frac{46}{1729} \approx .0266.$$

The Chicago Wheel paid off at 100 for 1 if a winner's numbers all appeared in one column and 10 for 1 if they were split 2–1 between the columns. The probability of all 3 numbers landing in a single column can be computed as a fraction of the probability p computed above. Once the 3 numbers are selected into a column, we need to count the number of ways to pick the other 9 numbers in that column from the 24 drawn numbers. We have

$$P(\text{3–0 split}) = p \cdot \frac{\binom{21}{9}}{\binom{24}{12}} = \frac{5}{1729}.$$

By the Complement Rule, the probability of a 2–1 split between the columns is then $\frac{41}{1729}$, since we are dividing the winning probability p into the probability of 2 disjoint events. The expected value of a 1-unit Chicago Wheel ticket is

$$E = (100) \cdot \frac{5}{1729} + (10) \cdot \frac{41}{1729} - 1 \approx -.4737,$$

and so the operator held a 47.4% edge—comparable to a state lottery ticket. Some other policy wagers are considered in Exercises 3.9–3.13.

Baccarat

Baccarat is a card game similar to blackjack that is often played in the high-limit area of casinos and is favored by many high rollers, among whom it is not unusual to risk tens of thousands of dollars on a single hand. Outside high-limit areas, it is seen in a scaled-down form known as *mini-baccarat* or "mini-bac" for short. In either case, the rules are simple: gamblers bet on which of two hands—called the *Player* and the *Banker*—will be closer to 9. These bets are paid at 1 to 1. A third option is to bet that the two hands will tie, which pays off at 8 to 1. If the hand is a tie, bets on Banker and on Player are also ties, and no money is won or lost on them.

Two cards, from a shoe (a box that holds multiple decks of playing cards) containing six to eight decks, are initially dealt to each hand. The Riviera Casino in Las Vegas once used a 16-deck shoe in an effort to encourage players to gamble longer—many players leave almost reflexively at the conclusion of a shoe while the cards are being shuffled. The Union Plaza Casino in downtown Las Vegas (now the Plaza) experimented briefly with what was billed as "the world's largest baccarat shoe" (see Figure 3.4), which held 144 decks, or 7488 cards, but this novelty was short-lived [131]. With a single deck of cards

FIGURE 3.4: A $3 Union Plaza baccarat chip advertising their 144-deck shoe.

measuring about 1.6 centimeters thick, this monster shoe measured about 7.5 feet long, held nearly 30 pounds of cards, and would certainly have been a challenge for the dealers to use.

Hand values are computed by counting each card at its face value, with aces counting 1 and face cards 0. If the sum of the cards exceeds 9, then the tens digit is dropped. For example, a hand consisting of a 7 and a 5 has value 2. Unlike in blackjack, a baccarat hand does not "bust" if it exceeds the highest possible hand value of 9.

If either hand has a value of 8 or 9, the hand is called a *natural*, and no further cards are drawn. If there are no naturals, an intricate set of rules dictates when either hand may receive a third card. The Player hand, which always goes first, draws a third card if its value is 5 or less. If the Player stands with 6 or 7, the Banker hand draws a third card if its value is 5 or less. If the Player draws a third card, the Banker hand's action is determined by its value and the value of the Player's third card. Table 3.11 shows the standard set of rules.

TABLE 3.11: Baccarat rules for the Banker hand [86]

Banker's hand	Banker draws if Player's 3rd card is
0–2	Anything
3	Not an 8
4	2–7
5	4–7
6	6 or 7
7	None

Example 3.2.21. Suppose that the cards dealt to the Player hand are $K\heartsuit$ and $7\diamondsuit$, and that the Banker hand receives $7\spadesuit$ and $6\diamondsuit$. The Player hand totals 7, and so does not draw a third card. Since the Banker's total is 3, which is less than 5, a third card is dealt to the Banker hand. If this card is the $Q\heartsuit$, the hand remains a 3, and so the Player wins, 7 to 3. ∎

Example 3.2.22. In the next hand, suppose the Player's hand is $9\heartsuit$ and $2\heartsuit$, for a total of 1, and the banker is dealt $K\spadesuit$ and $4\heartsuit$, totaling 4. The Player hand draws a third card, the $6\spadesuit$, for a total of 7. Since the Player's third card was a 6, a Banker hand of 4 is required to draw a third card. If that card is the $6\clubsuit$, the Banker hand is now 0, and Player wins, 7 to 0. ∎

These rules include some cases where the Banker hand is required to draw another card even though it's beating the Player hand already. For example, suppose that the Player hand is A-3, totaling 4. The Player takes a third card and draws a 7, bringing that hand's total to 1. If the Banker's hand is 9-6, for a total of 5, the Banker leads but must nonetheless take another card since the Player's third card was a 7. It is also possible for the Banker hand to be denied a third card even if it's tied with the Player hand, as in the case where the Player has 2-3 and draws an 8 for a total of 3, and the Banker holds 3-K. Because the Banker has 3 and the Player's third card was an 8, no further card will be drawn to the Banker hand, and the round ends in a tie.

How did game designers settle on these rules? Consider the rule cited above for drawing a third card to a Banker hand of 3. If the Player hand has drawn an 8 as its third card, it follows that the hand's value is 8, 9, 0, 1, 2, or 3, according as the initial two-card hand was 0, 1, 2, 3, 4, or 5. We may assume that these six totals occur equally often in the long run. If the Banker hand is 3, it wins half the time and ties one-sixth of the time, and the desire for a balanced game makes this a good place to stop drawing cards.

Since the rules of baccarat are fixed, it is entirely a game of chance. Contrary to the image of baccarat as presented in popular culture, such as when James Bond plays the game in six movies, skill plays no part in the play of the game. The replacement of baccarat by Texas hold'em in the 2006 version of *Casino Royale* represents something of a betrayal of the Bond legend even as it tapped into a wave of poker popularity.

In light of these complicated rules and the principle that "the easier a bet is to understand, the higher the house advantage," we might expect baccarat to have a fairly low HA, and indeed this is so. In an eight-deck game, the probability of winning a resolved bet (i.e., when the hand is not a tie) on Player is .4932 and the probability of winning a resolved bet on Banker is .5068 [86]. As with the Don't Pass bet in craps, it is not in the casino's best interests to offer a bet with a positive expected value, and so it is standard practice for casinos to charge a 5% commission on winning Banker bets. In effect, this means that if you win a bet on Player, you are paid at 1 to 1 odds, but if you bet on Banker and win, you receive only 95% of the amount wagered (19 to 20 odds).

As a convenience to everyone involved, players and dealers alike, some casinos use special $20 chips at their baccarat tables; this is a denomination not typically used elsewhere in the casino. These chips make the calculation of commissions easy: $1 for every $20 chip in play on a winning Banker wager. In practice, to avoid complicated change-making transactions, many casinos pay all winning Banker bets at even money, and commissions that an individual gambler owes are simply counted during the play of a shoe. When a player wishes to leave or the round has ended and the cards are being shuffled or replaced, the casino collects the accumulated commissions owed.

Example 3.2.23. As is also the case in blackjack, the probabilities of winning baccarat wagers depend on the number of decks of cards being used. If eight decks are used, these rules lead to the following probability distributions:

- Let X be the return on a $1 Player bet. The probability distribution for X is

x	-1	1	0
$P(X = x)$	.4584	.4461	.0955

and the corresponding expectation is

$$E(X) = (-1) \cdot .4584 + (1) \cdot .4461 + (0) \cdot .0955 = -\$.0123,$$

so the house advantage is 1.23%.

- If Y is the return on a $1 Banker bet, the distribution for Y is

y	-1	$.95$	0
$P(Y = y)$	.4461	.4584	.0955

and the expected value is

$$E(Y) = (-1) \cdot .4461 + (.95) \cdot .4584 + (0) \cdot .0955 = -\$.0106,$$

for an HA of 1.06%.

- Finally, if Z is the return from a \$1 Tie bet, the payoff is 8 to 1, and the distribution is

z	-1	8
$P(Z = z)$	.9045	.0955

The resulting expected value is then

$$E(Z) = (-1) \cdot .9045 + (8) \cdot .0955 = -\$.1405,$$

for an HA of 14.05%—this is clearly a bet to be avoided.

∎

As a promotional tool intended to attract bettors to the baccarat tables, casinos have been known to experiment with lowering the commission charged on winning Banker bets, which, of course, reduces the HA. If the commission is reduced to 4%, the house advantage drops to .60%, and at a 3% commission, the HA is a nearly invisible .14%—the game is essentially even at that level, as a \$100 bet loses, on the average, only 14¢.

In considering the mathematical aspects of baccarat, we are immediately drawn to one difference between it and other games we have considered: successive hands are not independent. As is also the case in blackjack, a card dealt in one hand is a card that cannot be dealt in the next hand, and so the probabilities of different hand totals change as the dealer moves through the shoe. That being the case, a mathematical analysis of baccarat requires somewhat more involved calculations that account for the changing deck composition. The following extended example will illustrate this.

Example 3.2.24. A variation of the baccarat stand or draw rules gives the gambler making the largest Player wager the option of deciding whether a Player hand of 5 draws a third card or stands. This less common version, sometimes called *chemin de fer*, appears to have a skill component, but does it really, or is the casino offering the player a meaningless choice?

That question can be answered with some simple reasoning, and aside from accounting for the lack of independence between hands, requires very little mathematics. Assuming that the Banker hand is not a natural, a Player hand of 5 wins against Banker hands of 0 to 4, ties a Banker 5, and loses only against a 6 or 7. The player making this decision has full knowledge of the Banker hand and a large bet on Player, and so certainly should choose to draw when losing to a 6 or 7—as the standard baccarat rules provide. In chemin de fer, the Banker hand will not draw to a 6 or 7 if the Player hand does not take a third card, and so will win if the player chooses not to draw.

Against a Banker hand of 0 to 4, the player in control of the decision must weigh the possibility of improvement against the fact that the hand is winning as dealt. If the player chooses never to draw to a 5, the Banker hand will draw another card, and so has a chance to convert a losing or tying hand into a winning hand. There is no instance where a winning Banker hand is converted to a nonwinning hand when the player stands on 5—at worst, a losing Banker hand remains a loser; it might improve and tie or win. It follows that standing on a 5 in these circumstances does not improve the Player hand's chance of winning.

If the hands are tied and the Player hand stands on 5, then the Banker hand will draw a third card. A decision by the player not to draw to the Player hand is a gamble that the next card out of the shoe will not improve a 5. While the exact probability of improvement depends on the cards that have been dealt, including the exact cards that constitute the two dealt 5s, we can get a sufficiently accurate estimate of the probability by assuming a full shoe. If the player chooses to draw a third card to a 5:

- The hand is improved if the third card dealt to it is an ace through 4, and the probability of improvement is therefore $\frac{20}{52} = \frac{4}{13}$. If that third card is not a 4, then Table 3.11 indicates that the Banker hand does not draw again, and so the Player wins with probability $\frac{3}{13}$.

- The hand stays the same if a face card or 10 is drawn, an event which also has probability $\frac{4}{12}$. The Banker hand does not draw under these circumstances, and so the hand remains tied.

- The hand loses value—dropping below 5—on any other card: a 5 through 9. This has probability $\frac{5}{13}$. If the third card is an 8 or 9, then the Banker hand does not draw a third card, and so wins automatically. On a 5, 6, or 7 (final Player hand of 0, 1, or 2), the Banker must draw again.

We have the following probabilities:

Event	Probability
Player wins	$3/13$
Banker wins	$2/13$
Hand is a tie	$4/13$
Banker draws	$4/13$

We shall break down the four subcases, each with probability $\frac{1}{13}$, that make up the "Banker draws" outcome. If the Player's third card is a 4, then the Banker hand is drawing against a Player 9, and so cannot win. The Banker ties only if the third card drawn to that hand is also a 4. The probability of drawing a 4 depends on how many 4s are included among the five cards already played; we know that that number is at least 1. An upper bound for the probability of a tie, assuming an eight-deck shoe, is $\frac{31}{411} \approx .0754$, since there are no more than 31 4s left among the 411 undealt cards. The maximum number of 4s removed from the deck is 3, in the case where each hand has been dealt a 4 and an ace initially, and so the probability of a tie is no less than $\frac{29}{411} \approx .0706$. (.0706–.0754 will also be the range of the probability of a tie in the other situations where the Banker must draw, because in each case we are simply computing the probability that the Banker's third card matches the Player's third card.) The probability that the Player hand wins is then between $\frac{380}{411} \approx .9246$ and $\frac{382}{411} \approx .9294$.

In the other cases where the Banker hand must draw a third card, the Banker can win or tie, and can lose if the Player hand is not 0. The ranges of the various probabilities are collected in Table 3.12.

TABLE 3.12: Chemin de fer: Probabilities when Player draws to a 5 and Banker also draws

| Player | | Outcome | | |
Third card	Total	P(Banker win)	P(Player win)	P(Tie)
4	9	0	.9246–.9294	.0706–.0754
5	0	.9246–.9294	0	.0706–.0754
6	1	.8467–.8540	.0730–.0779	.0706–.0754
7	2	.7689–.7762	.1509–.1557	.0706–.0754

The "Player draws 4" and "Player draws 5" cases are mirror images of one another; in the other two cases, the Banker has a large edge over the Player. How does this edge compare to the edge of $\frac{1}{13}$ that the Player has in all other cases?

We shall consider the case where neither the Banker nor the Player hands contains a card of the same rank as the Player's third hand. Under this assumption, the probability of a Banker win is

$$p_B = \left(\frac{1}{13}\right) \cdot (.9294 + .8540 + .7762) = .1969,$$

the probability of a Player win is

$$p_P = \left(\frac{1}{13}\right) \cdot (.9294 + .0779 + .1557) = .0895,$$

and the probability of a tie is

$$p_T = \left(\frac{4}{13}\right) \cdot .0706 = .0217.$$

Combining this information with the probabilities in the first three lines of the table above gives the following probabilities of the three possible outcomes when the player chooses to draw to a 5 against a Banker 5.

Event	Probability
Player wins	$3/13 + .0895 \approx .3202$
Banker wins	$2/13 + .1969 \approx .3507$
Hand is a tie	$4/13 + .0217 \approx .3293$

As a result, the expectation of a $1 bet in this situation is

$$E = (1) \cdot .320 + (-1) \cdot .351 + (0) \cdot .329 = -\$.031.$$

This value must be compared to the expected value when the player stands on 5. There are three possible outcomes.

- *The Player wins despite the Banker's draw.* This happens when the third card dealt to the Banker hand is a 5 through 9. The number of such cards left in the shoe is between 156 (if, for example, both Player and Banker hands are an 8 and a 7) and 160 (as is the case when both hands consist of an ace and a 4, or a 2 and a 3). Since only four cards have been dealt, 412 remain in the shoe, and so

$$\frac{156}{412} \leqslant P(\text{Player win}) \leqslant \frac{160}{412}.$$

- *The hand remains tied.* The Banker hand remains a 5 if its third card is a 10 or a face card. There are between 126 and 128 10-count cards left in the shoe, as each of the Player and Banker hands can have at most 1 such card. It follows that

$$\frac{126}{412} \leqslant P(\text{Tie}) \leqslant \frac{128}{412}.$$

- *The Banker hand draws a third card and wins.* The Banker's third card must be in the range ace–4. There are 124–128 such cards remaining, because the two hands dealt may contain, between them, 0, 2, or 4 of these low cards, and thus

$$\frac{124}{412} \leqslant P(\text{Banker win}) \leqslant \frac{128}{412}.$$

We can see immediately that at the start of a shoe, the probability that the Player wins when standing on 5 against a Banker 5 exceeds the probability of a Banker win. The Player's expectation is therefore positive when standing on 5 at the start of a shoe, and so standing has a higher expectation than drawing against a 5. Absent any useful information about the composition of the remaining cards, we have the following strategy for the lead bettor when the Player hand is dealt a 5:

Banker hand	Action
0–4	Draw
5	Stand
6–7	Draw

Since an optimal strategy exists, any deviation from this strategy increases the casino's advantage, and this may be a reason why a casino would offer this choice to a player—remember that the choice is given to the bettor making the largest Player bet, and so a choice that decreases the Player hand's chance of winning increases the likelihood that the casino will collect that largest wager. ∎

Texas Hold'Em

Texas hold'em is arguably the most popular form of poker in the 21st century. Though the game dates back to well before 2001, its recent rise in popularity can be attributed to televised tournaments which use small cameras to reveal players' hole cards to a viewing audience. Additionally, televised hold'em games use computers to calculate and display the probability of each player's hand winning the pot by rapidly running through each possible finished game in real time and identifying the best hand—an exercise in experimental probability that enhances the experience for the viewers.

The best possible hole cards are a pair of aces. The probability of drawing a pair of aces was computed on page 41 and found to be $\frac{1}{221}$. At the other end of the spectrum, the worst hole cards are a 2 and 7 of different suits: two low-ranked cards that are too far apart to fill in a straight and with no start on drawing a flush. The probability of an offsuit 2–7 is

$$\frac{4 \cdot 4 - 4}{\binom{52}{2}} = \frac{12}{1326} = \frac{2}{221}$$

—twice as likely as a pair of aces.

Example 3.2.25. Suppose that your hole cards are $A\heartsuit\ T\heartsuit$. Find the probability that the board completes a flush.

There are 11 hearts left unseen, and completing a flush requires 3 of them, but it doesn't matter where they fall among the flop, turn, and river. The probability of getting 3 or more hearts in 5 cards is

$$\frac{\binom{11}{3} \cdot \binom{39}{2} + \binom{11}{4} \cdot \binom{39}{1} + \binom{11}{5}}{\binom{50}{5}} = \frac{135,597}{2,118,760} \approx .0640,$$

or about 6.4%.

This includes the small probability that 5 hearts appear on the board and every player holds at least a flush, but in that circumstance, you hold the best, or *nut*, flush since the $A\heartsuit$ is in your hand. Only a straight flush can beat you if there are 5 hearts on the board. Since you hold the $A\heartsuit$ and $T\heartsuit$, only 4 straight flushes are even possible in other player's hands.

If only one heart falls on the flop, it may be advisable to fold, but if you stay in, what is the probability of completing the flush on the turn and river?

Here, there are 10 hearts left, but you need 2 of them in 2 cards. This probability is

$$\frac{\binom{10}{2}}{\binom{47}{2}} = \frac{10 \cdot 9}{47 \cdot 46} \approx .0416,$$

roughly two-thirds of the probability initially computed. ∎

The probability of a royal flush in a 5-card game is 1 in 649,740. With 7 cards to choose 5 from, but far more 7-card combinations than 5-card ones, how does this probability change in Texas hold'em?

There are 4 royal flushes; each may be combined with any 2 of the other 47 cards in the deck. The probability of a royal flush is

$$\frac{4 \cdot \binom{47}{2}}{\binom{52}{7}} = \frac{1}{30,940}.$$

Royal flushes may be 21 times more likely, but there is a small chance,

$$p = \frac{4}{2,598,960},$$

that the 5 cards of the royal flush all appear on the board, generating a tie among all players who have not folded.

Counting Outs and the Rule of Four

When assessing a hold'em hand after the flop, a player knows 5 cards. If that hand holds 4 cards to a flush, the probability of completing the flush on the turn is $\frac{9}{47} \approx .1915$, and the probability of missing the flush on the turn but filling it in on the river is

$$\frac{38}{47} \cdot \frac{9}{46} \approx .1582,$$

giving an overall probability of making the flush of about 34.97%.

To complete this calculation in one step, we use the Complement Rule. Suppose that the hand contains 4 clubs.

$$P(\text{Flush}) = 1 - P(\text{Draw 0 clubs}) = 1 - \frac{\binom{38}{2}}{\binom{47}{2}} \approx .3497.$$

The cards that will complete a hand are known as *outs*; in this example, there are 9 outs to the hand. More generally, if a hand has k outs, the probability P_k of catching 1 or 2 cards and completing the hand is

$$P_k = 1 - \frac{\binom{47-k}{2}}{\binom{47}{2}} = \frac{93k - k^2}{2162}.$$

This formula is too complicated to evaluate on the fly at a poker table with limited time to think, so an excellent approximation called the *Rule of Four* can be used instead [40].

Theorem 3.2.2. *The Rule of Four: The probability P_k of making a hand on the turn or river when there are k outs is approximately $4k\%$, or $\dfrac{4k}{100}$.*

Before proving this result, we shall illustrate it with an example. When drawing to an open-ended straight such as $89TJ$, which can be completed to a straight by 8 cards, we have $P_k \approx .3145$. The Rule of Four estimates this probability as .3200, which is less than 1.8% from the true value.

Proof. We shall consider this proposed approximation both algebraically and graphically as evidence for its accuracy.

We seek to approximate

$$P_k = \frac{93k - k^2}{2162}.$$

Factoring in the numerator gives

$$P_k = \frac{(93 - k) \cdot k}{2162}.$$

The Rule of Four claims that

$$\frac{93 - k}{2162} \approx \frac{4}{100} = .04.$$

The left side of this approximation is a decreasing function of k which is, to 4 decimal places, .0430 at $k = 0$ and falls to .0375 at $k = 12$. Over the interval $0 \leqslant k \leqslant 12$, then, the proposed approximation is within .0030 of .04; acceptably close to establish the desired result.

Figure 3.5 shows a graph of the difference $P_k - \dfrac{4k}{100}$ for $0 \leqslant k \leqslant 12$. P_k is greater than the Rule of Four estimate until the two functions are equal, at $k = 6.52$. The difference between the two functions over this interval is seen to be small, further confirming that the approximation an accurate one. If $k \leqslant 14$, the Rule of Four gives an approximation that is within 10% of P_k, and this is close enough to be useful in gameplay.

$\square$

FIGURE 3.5: Error in the Rule of Four approximation: $P_k - \dfrac{4k}{100}$.

Table 3.13 illustrates the accuracy of the Rule of Four by comparing P_k and $\dfrac{4k}{100}$ numerically.

TABLE 3.13: Probability P_k of completing a hold'em hand with k outs and Rule of Four approximation $4k/100$

k	P_k	$4k/100$
1	.0426	.0400
2	.0842	.0800
3	.1249	.1200
4	.1647	.1600
5	.2035	.2000
6	.2414	.2400
7	.2785	.2800
8	.3145	.3200
9	.3487	.3600
10	.3839	.4000
11	.4172	.4400
12	.4486	.4800
13	.4810	.5200
14	.5116	.5600

For a better approximation when $11 \leqslant k \leqslant 18$, use

$$P_k \approx \frac{4(k-1)}{100}.$$

A similar rule, suitable for estimating the probability of making a hand on a one-card draw, is the *Rule of Two*. In considering a one-card draw—the turn or the river—to a hand with k outs, this rule states that the approximate percent chance of successfully completing the hand is twice the number of outs.

For a 1-card draw with k outs from an n-card deck, the Rule of Two states that

$$\frac{\binom{k}{1}}{\binom{n}{1}} = \frac{k}{n} \approx \frac{2k}{100},$$

which is equivalent to saying that $n \approx 50$. Since the number of unknown cards on the turn is 47 and the number on the river is 46, this approximation is accurate. Multiplying by 2 is far easier than dividing by 47 or 46.

Bad Beat Bonus

A *bad beat* in hold'em is when a strong hand—defined, in some card rooms, to be aces full or higher—is beaten at the showdown by an even higher hand.

Example 3.2.26. Suppose that Robin's hole cards are $A\heartsuit\ A\clubsuit$ while Terry holds $K\diamondsuit\ Q\diamondsuit$. If the flop comes $A\spadesuit J\diamondsuit\ T\diamondsuit$, then Robin's hand is three aces and Terry holds a 4-card straight flush with a possible royal flush draw. If the turn and river are the $J\clubsuit$ and $9\diamondsuit$, Robin improves to aces full of jacks, but loses to Terry's straight flush.   ■

In an effort to cushion some of the blow of a crushing disappointment when a very high hand—well into the top 1% of all 5-card hands—nonetheless falls to an opponent, some card rooms offer a Bad Beat Bonus or Bad Beat Jackpot payoff to a player whose hand is on the losing end of such a showdown.

Boulder Station, Red Rock, and Santa Fe Station, three casinos operated by Las Vegas-based Red Rock Resorts, collaborate remotely on a Jumbo Hold'em Jackpot promotion that pays a progressive jackpot on bad beats. The progressive jackpot starts at $75,000 and increases once a day by 1–10% of the jackpot drop (an amount taken from each pot to fund casino promotions) across all three participating poker rooms. As the jackpot rises, the rules for awarding the jackpot change. The $75,000 starting value requires that a hand of $AAAJJ$ or better lose to four-of-a-kind or higher on the flop, while a jackpot of $100,000 or higher can be won by $AAAJJ$ or better losing to four of a kind at any point during the hand.

The progressive jackpot is divided as follows:

- 40% goes to the holder of the beaten hand.
- 25% is awarded to the winner of that hand.

  If 2 or more high hands lose on the same hand, the winner's share is reduced and the losing players' share is divided among the losing hands, with the highest losing hand receiving a larger share.

- 3% is divided evenly among all of the players at the table where the jackpot was won, including the winner and loser.
- The remaining 32% is divided among all active hold'em players at the 3 participating poker rooms when the jackpot was hit, including the players at the decisive table who shared 3% of the jackpot earlier.

One scenario that often leads to bad beats is when 3 aces fall on the flop; this has probability

$$\frac{\binom{4}{3}}{\binom{52}{3}} = \frac{1}{5525}.$$

Any player who played a pair in the hole now has aces full, but they will lose to a player who holds the fourth ace among their hole cards and cannot catch up by drawing into a straight flush, since they hold cards of 2 ranks with only the turn and river to come.

The probability that the board contains exactly 3 aces, which might trigger a bad beat bonus even if the aces aren't all part of the flop, is

$$\frac{\binom{4}{3} \cdot \binom{48}{2}}{\binom{52}{5}} \approx \frac{1}{576}.$$

In the Red Rock Resorts promotion described above, a player whose pair in the hole is completed to four-of-a-kind by the turn and river could claim the loser's share of the prize pool by losing to 4 aces—provided that the jackpot is at least $100,000.

Example 3.2.27. Jackpots of this size are only awarded when low-probability events happen. How low is "low"?

Suppose you are playing in a 6-handed game. If your hole cards are 9♣ 9♡ and 3 aces come on the flop ($p = \frac{1}{5525}$; above), the probability that the turn and the river are then the 9♠ and 9♢, in either order, is

$$\frac{1}{\binom{47}{2}} = \frac{1}{1081}.$$

If this happens, the probability that one of your 5 opponents holds the fourth ace and beats your four 9s with four aces is

$$\frac{\binom{44}{9}}{\binom{45}{10}} = \frac{2}{9}.$$

If this sequence of 3 improbable events happens *and* the jackpot is at least $100,000, you qualify for the loser's share of the jackpot. The exact amount of your win depends on how many players are at the 3 poker rooms involved, but it'll be at least $40,000. ∎

At the Mandalay Bay poker room on the Las Vegas Strip, the Aces Cracked promotion in the fall of 2021 offered a $100 consolation prize to any player whose pocket aces wound up as a losing hand in a pot amounting to at least $10. This prize was awarded up to 4 times during each gaming session, which ran from 3:00 to 6:00 P.M., Thursday through Monday.

Bad beat jackpots of this size invite collusion among players, since the prize for winning a $30 pot by cracking an opponent's aces is far less than the $100 jackpot paid to the losing player. Players in such a game might share information about their hands and agree to split the proceeds of the bonus, which is contrary to the "one player, one hand" rule that is standard in card rooms. The official rules of both Aces Cracked and the Jumbo Hold'Em Jackpot explicitly stated that any player discussion of hands during play could forfeit eligibility for this jackpot.

Video Poker

While Texas hold'em is the version of poker that has come to dominate both live play in casinos and televised play in poker tournaments, *video poker* is primarily a game of five-card draw poker simulated by a computer (Figure 3.6). As with slot machines, the player is playing against a fixed pay table

FIGURE 3.6: Video poker machines at Harrah's Casino, New Orleans, Louisiana [76].

rather than another human or electronic opponent. The goal is simply to end up with the best possible five-card poker hand: five cards are dealt, and the player may discard any or all of them and receive replacements.

There are two places in playing video poker where skill plays a part. The first, which begins before any money is put down, is selecting the best machine—in terms of the pay table offered—to play. The second, of course, lies in knowing what to discard and which hands to pursue in the draw. We begin with the first. The expected return for a video poker machine is typically expressed as its *payback percentage*, which with the right pay table and perfect play on the part of the player, can very occasionally exceed 100%— that is, the machine is expected to return more money than it takes in, in the long run. For example, Table 3.14 is a video poker pay table where all of the payoffs are "for 1," and include the amount wagered, so the 1 for 1 payoff for a pair of jacks or better is simply a break-even payoff [59, p. 121]. Most video poker machines accept wagers of one to five times the minimum bet; the amount wagered is typically described in the pay table as "n coins" even if the minimum amount does not correspond to a physical coin, as is the case, for example, with a $5 video poker machine.

TABLE 3.14: Jacks or Better video poker pay table with 99.8% return

Poker hand	Payoff: 1 coin	Payoff: 5 coins
Royal flush	250	4700
Straight flush	50	250
Four of a kind	25	125
Full house	9	45
Flush	6	30
Straight	4	20
Three of a kind	3	15
Two pair	2	10
Pair of jacks or higher	1	5

It is customary to describe video poker pay tables by the payoff for a one-coin bet on a full house and a flush; thus this would be a *9/6* game. With all other payouts equal, a 9/6 machine is better than a 9/5 machine, which is in turn better than an 8/5 machine.

One thing to notice immediately is that, when betting five coins, the payoff on every hand but one is simply five times the payoff with a single coin at risk. The exception—and it's a big one—is the royal flush. This is important, since approximately 2% of the payback percentage on most video poker machines comes from royal flushes betting the maximum, or playing "max coins" for short [142]. By not ruthlessly pursuing royal flushes whenever the strategy tables advocate doing so, the player is giving up on a meaningful part of the payoff: reducing a 98% machine to a 96% one, for example. This means that not only should a video poker player bet max coins, but also that the drawing strategy should be tilted toward pursuing royal flushes, which have a probability of approximately 1 in 40,000 on a video poker machine. Contrast this with the probability of 1 in 649,740 of receiving a *dealt* royal flush, and we can see the value of being allowed to discard cards and replace them in the draw.

How do we arrive at $\dfrac{1}{40,000}$ as the approximate probability of a royal flush? Consider 3 ways to get a royal flush at video poker:

- The probability of a dealt royal flush is the same as it is in any 5-card poker game:

$$P(\text{Dealt royal flush}) = \frac{4}{\binom{52}{5}}.$$

- Another way to get a royal flush is to draw 1 card to an initial 4-card royal and get the fifth card. This event has a probability of

$$\frac{20 \cdot 47}{\binom{52}{5}} \cdot \frac{1}{47} = \frac{20}{\binom{52}{5}}.$$

- The probability of a royal flush when drawing 2 cards to a dealt 3-card royal is

$$\frac{4 \cdot \binom{5}{3} \cdot \binom{47}{2}}{\binom{52}{5}} \cdot \frac{1}{\binom{47}{2}} = \frac{40}{\binom{52}{5}}.$$

We might reasonably continue in this vein, considering 3-card draws to 2-card royals, but this ignores the reality that a hand such as $A\spadesuit\ Q\spadesuit\ 9\clubsuit\ 9\heartsuit\ 5\diamondsuit$ is best played by drawing 3 cards to the pair of 9s rather than pursuing a royal flush, so many 2-card royals will be discarded as the player tries to improve the starting hand differently.

Adding these 3 probabilities gives

$$P(\text{Royal flush}) \approx \frac{64}{\binom{52}{5}} = \frac{1}{40,608.75},$$

which is within 1.5% of $\dfrac{1}{40,000}$.

We have also omitted some lower-probability events that might be played out, such as receiving a royal flush on the draw after discarding all 5 initial cards. A video poker hand where all 5 cards should be discarded is sometimes called a *razgu*; its probability is about 3.5% [34]. Assuming that the razgu contains no 10s, the probability of drawing a royal flush after discarding a razgu is

$$\frac{4}{\binom{47}{5}},$$

making the probability of receiving a 10-free razgu and then drawing a royal flush approximately

$$(.035) \cdot \frac{4}{\binom{47}{5}} \approx \frac{1}{10,959,707}.$$

If an initial Jacks or Better hand contains a single jack, queen, king, or ace, with no straight or flush possibility, optimal strategy calls for the player to hold that card and draw 4 new cards, so no razgu can contain a card higher than a 10.

Omitting this and similar very small terms does not change the fact that the common estimate $P(\text{Royal flush}) \approx \frac{1}{40,000} = 2.5 \times 10^{-5}$ is very accurate. Using this as the probability of a royal flush allows some simple calculations about the frequency of royals in actual gameplay—for video poker machines do not know the laws of probability and so do not understand that they're supposed to give the skilled player a royal flush every 40,000 hands.

The payoff percentage for Table 3.14 is 99.8%, assuming perfect play. What is perfect play? It depends on the machine and on the pay table; part of intelligent video poker play is learning the strategy appropriate to your machine. The ideal strategy for a given machine and pay table has been determined through computer simulation of millions of video poker hands. For the pay table above, the strategy in Table 3.15 leads to that 99.8% payback percentage. The corresponding 9/5 machine's payback percentage is 98.8%, and an 8/5 machine that is otherwise the same pays back at 97.6% [59].

Table 3.15 should be read from the top down until you reach a combination contained in the hand you hold, and then you should discard appropriately and draw the number of cards indicated.

The following terms are used in the table:

- An *outside straight* is four cards comprising a straight that is open at both ends, such as 6789. Drawing either a 5 or a 10 will complete this straight.

- An *inside straight* is four cards comprising a straight, but with a "hole" in the middle, such as 4568 or 5689. Only a drawn 7 will complete any of these straights, so an inside straight has less potential for improvement, and thus is less valuable, than an outside straight.

  The term "inside straight" is also applied to straights of the form $A234$ and $JKQA$ which, due to the ace, are only open at one end.

- *High* cards, for the purpose of jacks or better video poker, are the jack, queen, king, and ace—those cards that, when paired, pay off.

At a number of places in the table (line 4 and especially line 12), one can see that the pursuit of a royal flush is prioritized over lesser payoffs. Occasionally, perfect strategy may call on the gambler to break up a winning combination in

TABLE 3.15: Video poker strategy for standard 9/6 Jacks or Better [59]

	When dealt	Draw
1:	Royal flush	0
2:	Straight flush	0
3:	Four-of-a-kind	0
4:	4-card royal flush	1
5:	Full house	0
6:	Flush	0
7:	Three-of-a-kind	2
8:	Straight	0
9:	2 pairs	1
10:	4-card straight flush	1
11:	High pair: J, Q, K, or A	3
12:	3-card royal flush	2
13:	4-card flush	1
14:	Low pair, 2–10	3
15:	4-card outside straight	1
16:	3-card straight flush	2
17:	Suited JQ, JK, or QK	3
18:	4-card inside straight (3–4 high cards)	1
19:	Suited JA, QA, or KA	3
20:	Nonsuited JQK	2
21:	Suited J10, Q10, or K10	3
22:	1 or 2 high cards	3–4
23:	5 mixed low cards (a razgu)	5

order to draw cards toward completing a royal flush. Unlikely as successfully completing a royal by drawing one or two cards may be, the high payoff at max coins leads to an expected value in excess of the hand being broken up.

Example 3.2.28. On a game played with Table 3.14 as pay table, if you are dealt

$$4\clubsuit\ 10\clubsuit\ J\clubsuit\ Q\clubsuit\ A\clubsuit,$$

then perfect video poker strategy dictates that you break up the flush by discarding the $4\clubsuit$ and draw one card, hoping to pull the $K\clubsuit$ needed to complete the royal flush. (In Table 3.15, line 4, 4-card royal flush, appears before line 6, Flush.) The greatly enhanced payoff for a royal flush with max coins wagered makes this the better play, as we can see by computing expected values. If you hold the $4\clubsuit$ and cash in the flush, your expected profit is $25 with a max coins bet, assuming a $1 coin. If you discard the 4 and go for the royal flush, the following outcomes are possible:

Result	Net Payoff	Probability
Royal flush	4695	$1/47$
Flush (nonroyal)	25	$7/47$
Straight	15	$3/47$
Pair (J, Q, or A)	0	$9/47$
Nonpaying hand	−5	$27/47$

The expected value of the hand if the 4♣ is discarded is

$$E = (4695) \cdot \frac{1}{47} + (25) \cdot \frac{7}{47} + (15) \cdot \frac{3}{47} + (0) \cdot \frac{9}{47} + (-5) \cdot \frac{27}{47}$$
$$= \frac{4780}{47} \approx \$101.70.$$

It follows that, even though the probability of winding up with nothing after the draw is over 50%, your expected return is over four times greater by giving up on the sure $25 payoff for the chance of hitting a royal flush. ∎

Some video poker machines modify this pay table by paying out 4000 coins for a 5-coin bet with a royal flush. The payback percentage on such a machine drops to 99.5%. Such a machine has an expectation of $86.81 for this hand, which is less than $101.70 but still better than the $25 expectation of holding the dealt flush.

Example 3.2.29. What should you do if you are dealt the following hand?

$$4\diamondsuit \ 4\clubsuit \ 5\clubsuit \ 6\heartsuit \ 7\heartsuit$$

The competing holdings here are a low pair (of 4s) and a 4-card outside straight. Reading down the chart, the pair appears on line 14 and the straight at position 15. It might be a close call, but the best play is to hold the pair. ∎

Low pairs, though they are not winning hands on the deal, nonetheless have value, and so they should be held rather than breaking them up in pursuit of an incomplete straight, or trying to pair a single dealt high card by discarding 4 cards.

Note that a three-card straight does not appear in the table, and thus is not worth holding unless it is suited (and hence a three-card straight flush; line 16) or consists of a jack, queen, and king (line 20). Similarly, the table never directs the player to hold a three-card flush unless it's part of a three-card royal or straight flush.

More significantly, the table never advises a player to hold a *kicker*: an unpaired high card held alongside a pair. If you have two 10s, a 5, a 2, and an ace of various suits, you should hold only the 10s and draw three cards, resisting the temptation to hold the ace as well, because holding onto it will not, in the long run, improve your position. This, of course, only applies to a game using Table 3.14 as its pay table. It is important when playing video

Basic Gambling Mathematics

poker to align your strategy with the rules and payoffs for the game you are playing. Nonetheless, holding a kicker is almost certainly going to decrease your long-term expectation, by as much as 4%, regardless of the exact game you're playing [59].

Consider a 3-card draw to $9\Diamond\ 9\clubsuit$, with $\binom{47}{3} = 16,215$ possibilities, compared to a 2-card draw to $9\Diamond\ 9\clubsuit\ A\Diamond$, which can be completed in $\binom{47}{2} = 1081$ ways. In either case, there is 1 rank (9s) with 2 cards remaining, 3 ranks with 3 cards remaining, and 8 ranks with all 4 cards still in the deck. The ways to improve these hands are compared in Table 3.16.

TABLE 3.16: Probabilities of the possible outcomes when drawing to a pair vs. a pair plus a kicker

Final hand	Draw 3	Draw 2
Four-of-a-kind	.0028	.0009
Full house	.0102	.0083
Three-of-a-kind	.1143	.0777
2 pairs	.1599	.1776
1 pair (no change)	.7129	.7354

Holding a kicker brings a slight increase in the probability of drawing into 2 pairs, but otherwise, the chance of improving is better without the third card, and the chance of no improvement and hence no payoff is greater when holding the ace.

Kickers may have some value in live draw poker when they might mislead opponents about the strength of your hand. When dealt a pair of 9s and an ace, holding the ace and drawing 2 cards represents your hand as possibly being three-of-a-kind or 3 cards to a straight flush. This illusion might influence your opponents to act in ways that benefit you, perhaps by folding or underbetting their hands. Since video poker is played against a fixed pay table, their is no opponent to outwit, and so no value in holding a kicker.

Example 3.2.30. What does the strategy chart say about the following hand?

$$5\clubsuit\ 8\heartsuit\ 4\Diamond\ 9\clubsuit\ 7\clubsuit$$

This hand contains the following:

- A three-card straight.
- A three-card inside straight flush.
- Two different four-card inside straights, although neither one contains any high cards.

Referring to Table 3.15, we see that line 16, "3-card straight flush," is the first one that matches our holdings, and so the correct course of action is to discard the 8$\heartsuit$ and 4$\diamondsuit$ and hope to complete the straight flush by drawing the 6$\clubsuit$ and 8$\clubsuit$. While the probability of pulling this off is only $\dfrac{1}{1081}$, it's the option with the highest expected value. ∎

It should be noted, of course, that there are multiple ways to win in this example; we might complete a flush or straight instead of a straight flush, or draw into two pairs or three of a kind.

The strategy table does not distinguish between inside and outside straight flushes because the payoff justifies the reach. If you hold three cards of a straight flush, the number of pairs of cards that will complete it ranges from one to three: 7$\spadesuit$ 8$\spadesuit$ 9$\spadesuit$ can be filled out to a straight flush with either the 5 and 6, 5 and 10, or 10 and jack of spades, while 4$\diamondsuit$ 6$\diamondsuit$ 8$\diamondsuit$ can only be completed to a straight flush with the 5 and 7 of diamonds. (A double inside straight flush such as 4$\diamondsuit$ 6$\diamondsuit$ 8$\diamondsuit$ is sometimes called a *kangaroo straight*.) In any event, it's worth drawing two cards in search of the 50 for 1 payoff.

By making some small modifications to the pay table, we can arrive at Table 3.17, which is an 8/8 game with a payback percentage of 101.1% with perfect play [59, p. 125].

TABLE 3.17: Video poker pay table with a 101.1% return

Poker Hand	Payoff: 1 coin	Payoff: 5 coins
Royal flush	250	4700
Straight flush	200	1000
Four of a kind	40	200
Full house	8	40
Flush	8	40
Straight	8	40
Three of a kind	3	15
Two pair	1	5
Pair of jacks or higher	1	5

Of course, "perfect play" on this machine would be different from perfect play on the previous pay table. In particular, the sharply increased payoff for straight flushes means that perfect play calls for a gambler to pursue those more aggressively, even to the point where "Two-card royal flush" and "Two-card straight flush," which have no standing in Table 3.15, appear as entries in the corresponding strategy table [59].

In Example 3.2.4, we considered a million-dollar sequential royal flush (SRF) promotion. Some video poker machines have added sequential royals as the highest-paying hand. If there is no significant reduction in the payoff for a non-sequential royal, then the expected return on the game increases. If this is coupled with a progressive jackpot awarded to the holder of a sequential royal, it is possible that the game may pay off more than 100%—again assuming perfect play, which would have to include different strategy to incorporate the pursuit of certain sequential royals. As a general guideline, altering video poker strategy to pursue sequential royals is only the superior play if you are dealt at least 2 cards of a sequential royal. Note that the queen must land in the center position in either order, so if a queen is dealt there, both SRFs remain possible on the draw.

However, one must be careful to inspect the pay table for changes to the payoffs on more common hands. *Bonus Poker* is a video poker variant that pays extra on some four-of-a-kind hands. These bonuses are paid for by lower payoffs for some lesser hands. Table 3.18 shows the pay table for a 5¢ Bonus Poker video game including a sequential royal bonus at the Aliante Casino in North Las Vegas, Nevada together with the pay table for a nearby Bonus Poker machine without the sequential royal bonus.

TABLE 3.18: Aliante Casino Bonus Poker pay tables

Poker Hand	SRF bonus		No SRF bonus	
	1 coin	5 coins	1 coin	5 coins
Sequential royal flush	250	Jackpot	250	4000
Royal flush	250	4000	250	4000
Straight flush	50	250	50	250
Four aces	80	400	80	400
Four 2s, 3s, or 4s	80	400	40	200
Four 5s through Ks	80	400	25	125
Full house	8	40	6	30
Flush	5	25	5	25
Straight	4	20	4	20
Three of a kind	3	15	3	15
Two pair	1	5	2	10
Pair of jacks or higher	1	5	1	5

The progressive jackpot at the time was $63,835.82, a handsome 255,341–1 payoff for a 25¢ bet.

Without the SRF bonus, the four-of-a-kind hands are lumped together and paid at the same rate, but the casino makes up for this largesse and funds the progressive jackpot by reducing the payoff on two pairs from 2 for 1 to a mere push. Players eschewing the SRF bonus take on the further burden of lower payoffs for full houses and some four of a kinds, but will make a 1-unit profit anytime they draw into 2 pairs.

Three Card Poker

If five-card poker hands are too complicated, perhaps restricting your hand to three cards might be more to your liking. Apparently enough people feel this way to make *Three Card Poker* (3CP) another carnival game that's found a place in the game lineup at a number of casinos.

The title gives the key to the game: players are dealt three cards apiece, and match their cards up against the dealer's hand—unlike Texas hold'em, this is a game where players face off against the dealer instead of each other. In 3CP, players make an Ante bet and then can choose, based on their cards, to fold or make a second "Play" bet. Play bets only have action if the dealer's hand "qualifies" by ranking queen-high or better. If the dealer fails to qualify, Ante bets pay even money and Play bets push. If the dealer qualifies, winning player hands pay even money on both the Ante and Play bets; if the dealer's hand is higher, then both bets lose. Additionally, the Ante bet pays a bonus if the player's hand is a straight or higher, even if it is beaten by the dealer.

The need for the dealer to reach a minimum qualifying hand raises an important question: What is the probability of the dealer qualifying? There are $\binom{52}{3} = 22,100$ three-card hands, and $\binom{40}{3} = 9880$ of them contain no queens, kings, or aces. From this latter number, of course, we must remove any hand containing a pair or higher.

For example: Any card from 4 to J can be the highest card in a three-card straight flush; these hands run from 432 through J109. There are 8 ranks and 4 suits to consider, for a total of 32 such hands.

These "exceptional" hands are counted in Table 3.19.

TABLE 3.19: Three Card Poker: Qualifying hands with no AKQ

Hand	Count
Straight flush, JT9 or lower	32
Three of a kind, jacks or lower	40
Flush, jack-high or lower	$4 \cdot \binom{10}{3} - 32 = 448$
Straight, JT9 or lower	$32 \cdot 4 \cdot 4 - 32 = 480$
Pair, jacks or lower	$10 \cdot \binom{4}{2} \cdot 36 = 2160$

Adding up in this table gives 3160 possible qualifying hands with no card higher than a jack, and subtracting gives 6720 nonqualifying hands. The probability that the dealer qualifies is therefore

$$p = 1 - \frac{6720}{22,100} = \frac{15,380}{22,100} \approx .6959,$$

so the dealer qualifies about 70% of the time.

The qualifying hand is the source of the house's advantage in 3CP. By requiring that the dealer hold at least a queen before Play bets have action, some winning hands will be underpaid, winning only the Ante bet instead of both the Ante and Play wagers.

Armed with this information, we look next at the pay table for player hands that beat a qualifying dealer hand. The Ante Bonus pay table varies among casinos; a common payoff structure is 1 to 1 on straights, 4 to 1 on three-of-a-kinds, and 5 to 1 on straight flushes [120]. Some casinos recognize a "Mini Royal Flush": a suited AKQ, and reward it with a 50 to 1 Ante Bonus payoff.

There is no pay table for the Play bet—all bets are paid at 1 to 1—but there is an optional Pair Plus bet that pays off player hands of at least a pair at odds. This bet must be made with the Ante bet before the cards are dealt, of course. Paytables can vary among casinos; one of the most common is Table 3.20. It should be noted that the Pair Plus bet is paid off even if the player's hand does not beat the dealer's or if the dealer fails to qualify.

TABLE 3.20: Three Card Poker payoffs: Pair Plus bet [120]

Hand	Payoff
Straight flush	40 to 1
Three of a kind	30 to 1
Straight	5 to 1
Flush	4 to 1
Pair	1 to 1

As we might expect, these payoffs are far below the probabilities of the various hands. In the case of a straight flush, there are 48 possible hands (any card except a deuce can be the high card in a three-card straight flush), and so $P(\text{Straight flush}) = \dfrac{48}{22,100} \approx \dfrac{1}{460}$.

In table games derived from poker where the player faces a decision to fold or make an additional bet after seeing some or all of her cards, optimal strategy suggests determining the minimum hand, the so-called *beacon hand*, for which the expectation is greater than the −$1 outcome achieved by folding and forfeiting the Ante bet. The beacon hand separating folding from playing on for 3CP turns out to be Q64: any player hand better than this should be backed with a Play bet and any lesser hand should be folded. Following this one piece of advice limits the casino's advantage to a mere 3.37%—actually quite reasonable for a table game. If instead you choose to "mimic the dealer" by calling on every hand holding at least a queen, that gives the casino a 3.45% edge—not much of a change. The nonstrategy of always calling, on the other hand, raises the HA to 7.65%: more than double that when using the beacon hand strategy above [120].

Sports Betting

When a bookmaker, legal or otherwise, sets odds on the two teams in a game, these odds are based on implied probabilities that each team will win. These "true" odds can be determined by finding the mean of the two sets of odds [39]. On page 69, we saw an example where bettors on the favorite in a contest had to give odds of 2–1 (or 10–5), betting \$2 to win \$1, while players backing the underdog received 8–5 odds. The true odds on the favorite were thus 9–5, meaning that the bookmaker had assessed the probability of the favorite winning as $\frac{9}{14}$.

It follows that the sports book has the edge on either wager. The expected value of bets on the favorite and the underdog are

$$E(\text{Favorite}) = (1) \cdot \frac{9}{14} + (-2) \cdot \frac{5}{14} = -\frac{1}{14}$$

and

$$E(\text{Underdog}) = (8) \cdot \frac{5}{14} + (-5)\frac{9}{14} = -\frac{5}{14}.$$

Both expectations are negative, so the bookmaker will extract a profit no matter which team wins. The HA is larger on the underdog bet; this might serve as some insurance to the bookmaker in the event that action will be unequal between the two teams.

Example 3.2.31. Consider a point spread bet on a single game. Assuming that the line has been properly set, players are essentially betting \$11 on a 50/50 proposition hoping to win \$10, and so the expectation is

$$E = (10) \cdot \frac{1}{2} + (-11) \cdot \frac{1}{2} = -\$\frac{1}{2},$$

which results in a house edge of approximately 4.55%, no matter which team the bettor chooses. If the point spread has been carefully chosen and attracts approximately equal action on each team, the sports book will make its 4.55% regardless of which team wins. ∎

In Example 2.4.3, we looked at the daily parlay cards offered by Station Casinos. If we assume that your chance of picking the winner of each game you bet on is ½ and that the games are independent, what is the expected value of a parlay card bet?

The expected value depends on how many games you select. In general, the expected value of a \$1 bet on an n-team parlay card is a function of n:

$$E(n) = x \cdot \left(\frac{1}{2}\right)^n - 1,$$

where x is the payoff, stated as "x for 1." Remember that when payoff odds are stated in this fashion, the payoff includes the amount of the original bet. For

TABLE 3.21: Parlay card payoffs and expectations: $p = .50$

n	Payoff	Expected value	HA
3	6 for 1	−$.25	25%
4	12 for 1	−$.25	25%
5	23 for 1	−$.281	28.1%
6	45 for 1	−$.297	29.7%
7	80 for 1	−$.375	37.5%
8	160 for 1	−$.375	37.5%
9	320 for 1	−$.375	37.5%
10	800 for 1	−$.219	21.9%

the payoff options offered on the parlay card, Table 3.21 shows the expected values and corresponding house advantages.

Some things are immediately apparent from this table:

- The house advantage on every bet offered is *huge*—on a par with keno and among the highest of all the bets available in a casino.

- The parlay with the lowest HA—picking ten games correctly—is also the one with the lowest probability of success, $\dfrac{1}{1024}$.

- The HA for picking 7, 8, or 9 games correctly is the same. This is because the payoff has doubled while the probability of winning has decreased by ½, and these factors cancel out.

- If the pattern for 7, 8, and 9 games had been continued through to 10 games, the payoff would have been 640 for 1. Since the 800 for 1 payoff is printed on the parlay card in larger and bolder type than the others, we may conclude that the casino wishes to attract action on this betting option by offering a higher payoff, and that, given the low probability of winning this bet, they do so at little risk.

But what if you know something about sports and can pick winners at a better-than-50% rate? If you can do even as well as 53% on every game, you can cut the HA to a far more reasonable level—and even turn the game to your advantage for $n = 9$ or 10. The EV for a $1 bet then increases to

$$E = x \cdot (.53)^n - 1.$$

These values, for the various values of n, are collected in Table 3.22. A negative HA in this table represents a player advantage.

Here, then, is one case where skill plays a part in gambling and where, under the right circumstances, the game can be turned in the gambler's favor.

TABLE 3.22: Parlay card payoffs and expectations: $p = .53$

n	Payoff	Expected value	HA
3	6 for 1	−$.1070	10.7%
4	12 for 1	−$.0530	5.3%
5	23 for 1	−$.0382	3.82%
6	45 for 1	−$.0026	.26%
7	80 for 1	−$.0602	6.02%
8	160 for 1	−$.0038	.38%
9	320 for 1	$.0559	−5.59%
10	800 for 1	$.3991	−39.91%

However, the chance of picking 10 out of 10 games correctly, even with a 53% chance of success per game, is still the low

$$\left(\frac{53}{100}\right)^{10} \approx \frac{1}{572}.$$

When betting a teaser card, though, 53% isn't enough to turn the expectation positive. Consider the 7-team, 7-point pro football teaser at the Golden Nugget (page 72), which pays 6–1. The expectation is positive if

$$E(p) = (6) \cdot p^7 + (-1) \cdot (1 - p^7) = 7p^7 - 1 > 0,$$

which requires

$$p > \sqrt[7]{\frac{1}{7}} \approx .7573,$$

just over ¾.

College football games tend to be somewhat more unpredictable than pro games, so the Golden Nugget offers slightly higher payoff odds on its college football teaser cards. A 7-team, 7-point college football teaser pays 7–1 instead of 6–1. This card has positive expectation if the gambler can pick winners, incorporating the 7-point bonus, with probability $p > .7430$, slightly less than ¾.

Neither level of accuracy is enough, however, to make an 8-team, 7-point teaser profitable at its 8–1 payoff. Positive expectation for that teaser requires that $p > \sqrt[8]{\frac{1}{9}} \approx .7598$.

Looking at a smaller ticket, consider the 2-team, 6½-point teaser offering a 2–3 payoff. The expected value is

$$E(p) = (2) \cdot p^2 + (-3) \cdot (1 - p^2) = 5p^2 - 3.$$

This requires $p > \sqrt{\frac{3}{5}} \approx .7746$ to be positive, so the situation is not any better on a 2-team teaser.

Repeating this calculation for all of the Golden Nugget's pro football teaser card options shows that the lowest probability yielding a positive expectation is the 6-team, 6-point teaser, with $p > .7230$, and the highest probability is required for the 4-team, 7-point card, for which $p > .7875$.

Reverse teasers turn that trend around. Lower chances of winning are countered by higher payoffs. The 6-team reverse teaser at Circa pays 750–1, and so has positive expectation when

$$E(p) = (750) \cdot p^6 + (-1) \cdot (1 - p^6) = 751p^6 - 1 > 0,$$

or when $p > .3317$. While it is true that picking winners against the shifted point spread requires less than 40% proficiency per game to be profitable, the gambler is playing against an unfavorable point spread, and it is quite possible, as we saw in Example 2.4.5, that reverse teaser point spreads can lead to neither team winning against its point spread.

Example 3.2.32. In early 2023, the sports book at the South Point Casino in Las Vegas was offering the following money line on the coin toss at Super Bowl LVII:

Heads: –102
Tails: –102

Translated into betting activity, this meant that, in making this bet, you would be risking $102 on the outcome of the coin toss, hoping to win $100 (or some other allowable wager at the same odds: $51 hoping to win $50 or $204 to win $200, for example). Since a coin toss is the ultimate 50/50 proposition, the expected value of this bet is

$$E = (100) \cdot \frac{1}{2} + (-102) \cdot \frac{1}{2} = -\$1,$$

and the corresponding house advantage is $\dfrac{1}{102} \approx .98\%$. ∎

While this is a very low HA, it cannot escape notice that making this bet calls for risking more money than you stand to win—on the literal toss of a coin. Similar –102 wagers both ways were available on whether the Kansas City Chiefs or Philadelphia Eagles would win the toss, and on whether the player who called the toss would be right or wrong. (The coin came up heads.)

Sports book manager Chris Andrews, who has worked at several casinos in Las Vegas and Reno, first offered a Super Bowl coin toss proposition at –120 either way. His intent was to point out how ridiculous betting on a coin toss was, but bettors flocked to this opportunity, even at an 8.33% disadvantage [3].

Punchboards

In considering the Ninety Percenter punchboard from Example 2.4.13, one is drawn to the fact that, due to the $5 prize for punching out the last spot, a gambler is guaranteed a profit if there are 49 or fewer unpunched spots: by purchasing and punching all of them. This profit is at least 10¢ and may be more if there are additional winning numbers still on the board. At the same time, purchasing all 800 spots on a fresh punchboard is a recipe for a guaranteed loss of $8 if the board is accurately named.

Given this progression from negative to positive expectation, it's reasonable to ask the following question: At which number of unpunched spots is the expected return from purchasing them all a maximum? To assess this question, we need to develop an expression for the expected number of $1-winning spots left on the board after a certain number of punches have been sold. Let the number of previous punches be n, where $0 \leqslant n \leqslant 799$, and denote by N the expected number of winners drawn; we then have

$$N = \sum_{k=0}^{n} k \cdot P(k \text{ winners in } n \text{ punches}) = \sum_{k=0}^{n} k \cdot \frac{\binom{67}{k} \cdot \binom{733}{n-k}}{\binom{800}{n}}.$$

N is a function only of n and may rightly be denoted $N(n)$ to show this functional relationship. Since $\binom{n}{r} = 0$ if $r > n$, this sum can be truncated with an upper limit of 67 if $n > 67$.

It follows that the expected number of $1 winners remaining after n punches is $67 - N(n)$. If you go on to purchase the last $800 - n$ spots, your expected return as a function of n is

$$E(n) = \$ \left(67 - N(n) + 5 - \frac{800 - n}{10} \right) = \$ \left(\frac{n}{10} - N(n) - 8 \right).$$

This can be simplified considerably:

$$E(n) = \frac{13}{800} n - 8,$$

from which we can see that the expectation is linear and increasing, and first exceeds 0 when $n = 493$. It follows that a Ninety Percenter punchboard with 307 or fewer spots remaining has a positive expectation.

The *maximum* expectation is achieved with one spot left to be punched, although it is difficult to imagine a situation where a gambler would have encountered this punchboard with only one spot remaining. Since the return on the last spot is guaranteed to be at least $5, it is unlikely that another gambler would have walked away when $5 (or possibly $6, with probability 67/800) could be had for a dime.

3.3 Binomial Distribution

We begin by considering an example:

> *Suppose you make 100 $1 bets on red at an American roulette wheel. What is the probability that you will win on 60 of the spins?*

We might consider this problem by looking at your total winnings, should this occur—that would be $20. Trying to calculate this probability directly quickly leads us into an algebraic morass from which it is difficult to emerge with an answer that we can trust.

Solving this problem is facilitated by introducing the concept of a *binomial experiment*.

Definition 3.3.1. A *binomial* experiment has the following four characteristics:

1. The experiment consists of a fixed number of successive trials, denoted by n.

2. Each trial has exactly two outcomes, denoted *success* and *failure*.

   In practice, it is often possible to amalgamate multiple outcomes into a single category to get down to two. For example, in the roulette problem posed above, we can denote the 18 red numbers as "success" and the 20 nonred numbers as "failure"—if we lose our bet on red, it matters little whether the number that was actually spun was black or green, or what the number was.

3. The probabilities of success and failure are constant from trial to trial. We denote the probability of success by p and the probability of failure by q, where $q = 1 - p$.

4. The trials are independent.

Definition 3.3.2. A random variable X that counts the number of successes of a binomial experiment is called a *binomial* random variable. The values n and p are called the *parameters* of X, and X is sometimes denoted as a $B(n, p)$ random variable to show dependence on the parameters.

The experiment described in the example above meets the four listed criteria and is therefore a binomial experiment. If we change the experiment to "Begin betting on red on successive spins of a roulette wheel, and let the random variable X be the number of spins required to win exactly 10 times," then the new experiment is not binomial. Since the number of trials is no longer fixed at the outset, criterion 1 is no longer true.

If X is a binomial random variable with parameters n and p, the formula for $P(X = r)$ can be derived through the following three-step process:

1. Select which r of the n trials are to be successes. This can be done in $\binom{n}{r}$ ways, as the order in which we select the successes does not matter.

   If we think of the trials as a row of n boxes, each to be designated "success" or "failure," what we're doing here is determining which r of the n boxes are successes.

2. Compute the probability of these r trials resulting in successes. Since the trials are independent, this probability is p^r.

3. We must now ensure that there are *only* p successes. This is done by assigning the outcome "failure" to the remaining $n - r$ trials. The probability of this many failures is $(1 - p)^{n-r} = q^{n-r}$.

Multiplying these three factors together gives the following result, called the *binomial formula*:

Theorem 3.3.1. *If X is a binomial random variable with parameters n and p, then*

$$P(X = r) = \binom{n}{r} \cdot p^r \cdot q^{n-r} = \binom{n}{r} \cdot p^r \cdot (1 - p)^{n-r}.$$

We can now revisit our original example. The experiment described meets the four criteria of a binomial experiment, where the amalgamation described in the definition is used to get down to two outcomes. The parameters of the experiment are $n = 100$ and $p = \dfrac{18}{38}$, and so with $r = 60$, we find that

$$P(X = 60) = \binom{100}{60} \cdot \left(\frac{18}{38}\right)^{60} \cdot \left(\frac{20}{38}\right)^{40} \approx .0033.$$

Binomial probabilities can be computed directly on a TI-84+ calculator by using the **binompdf** routine, which is found by pressing $\boxed{\textbf{2nd}}$ $\boxed{\textbf{DISTR}}$ and choosing option **A: binompdf**. The calculator will prompt you for the values of n, p, and r. "pdf" is an abbreviation for "probability distribution function."

Another calculator routine, **B: binomcdf**, computes binomial probabilities involving inequalities. "cdf" stands for "cumulative distribution function" and finds probabilities of the form $P(X \leqslant n)$.

Example 3.3.1. In the roulette betting scenario described above, what is the probability that you will be ahead after 100 bets?

Being ahead after 100 even-money bets corresponds to winning at least 51 bets, which may be interpreted probabilistically as $P(X \geqslant 51)$. Calculating this probability using the binomial formula would require 50 separate calculations and one large sum, or some advanced estimation techniques. The **binomcdf** routine, accessible as option **B: binomcdf** on the distribution menu, computes $P(0 \leqslant X \leqslant r)$ when provided with values for n, p, and r.

For this problem, we need $P(X \geqslant 51)$, which is $1 - P(0 \leqslant X \leqslant 50)$ by the Complement Rule. Using the TI-84+ shows that this probability is approximately .2650.

Suppose instead that you make 100 consecutive roulette bets on the number 13. What is the probability that you will be ahead after 100 spins?

A single-number roulette bet pays off at 35 to 1, so if you win three times, the $105 you win will offset your $97 in losses and leave you with $8 in profit. To show a profit, then, you must win at least three times, so we need to find $P(X \geqslant 3)$, or $1 - P(X \leqslant 2)$. With $n = 100$ and $p = \dfrac{1}{38}$, we compute this on the TI-84+ by finding $1 - P(X \leqslant 2)$, which is .1649. ■

If a random variable is binomial, computing its expected value is simple.

Theorem 3.3.2. *If X is a binomial random variable with parameters n and p, then $E(X) = np$.*

Put simply, the average number of successes is the number of trials multiplied by the probability of success on a single trial.

Proof.

$$E(X) = \sum_{x=0}^{n} x \cdot P(X = x)$$

$$= \sum_{x=0}^{n} x \cdot \binom{n}{x} \cdot p^x \cdot q^{n-x}$$

$$= \sum_{x=0}^{n} x \cdot \frac{n!}{(n-x)! \cdot x!} \cdot p^x \cdot q^{n-x}.$$

Since the $x = 0$ term is equal to zero, we can drop that term from the sum and renumber starting at 1:

$$E(X) = \sum_{x=1}^{n} x \cdot \frac{n!}{(n-x)! \cdot x!} \cdot p^x \cdot q^{n-x}$$

$$= \sum_{x=1}^{n} \frac{n!}{(n-x)! \cdot (x-1)!} \cdot p^x \cdot q^{n-x}$$

$$= np \cdot \sum_{x=1}^{n} \frac{(n-1)!}{(n-x)! \cdot (x-1)!} \cdot p^{x-1} \cdot q^{n-x}$$

$$= np \cdot \sum_{x=1}^{n} \frac{(n-1)!}{[(n-1) - (x-1)]! \cdot (x-1)!} \cdot p^{x-1} \cdot q^{n-1-x+1}.$$

If we substitute $y = x - 1$ in this last sum, we have

$$E(X) = np \cdot \sum_{y=0}^{n-1} \frac{(n-1)!}{[(n-1)-y]! \cdot y!} \cdot p^y \cdot q^{n-1-y}$$

$$= np \cdot \sum_{y=0}^{n-1} P(Y = y).$$

This sum is the sum of all of the probabilities of a binomial random variable Y with parameters $n-1$ and p, and so adds up to 1 (Axiom 2), completing the proof. □

Example 3.3.2. In Example 3.3.1, the expected number of wins in 100 spins of the wheel is $E = np = 100 \cdot \frac{18}{38} \approx 47.37$. ∎

It follows from this result that your average winnings after making 100 of these bets would be

$$(1) \cdot 47.37 + (-1) \cdot 52.63 = -\$5.26,$$

or 5.26% of the amount wagered—which is, of course, the long-term HA on a single bet.

Example 3.3.3. If we consider a bet on a single number rather than on red in American roulette (as in Example 3.3.1), the expected number of wins in 100 spins is $E = np = 100 \cdot \frac{1}{38} \approx 2.63$. Once again, your average winnings would be $-\$5.26$ after 100 \$1 bets. ∎

When betting a sports parlay card, the number of games that win is a binomial random variable X. If the probability of picking a game correctly is p, then the probability of picking k games on an n-game card correctly is

$$P(X = k) = \binom{n}{k} \cdot p^k \cdot (1-p)^{n-k}.$$

Of course, all of the games must win if the parlay card is to pay off. One might consider, instead of making a single parlay card bet, wagering on each game individually, at 10–11 odds against a point spread. While this requires a greater initial investment, the "all or nothing" factor is removed from the payoff. For n bets of \$11 each, the PDF for X, the number of wins, is the function $P(X = k)$ given above, but the total won or lost is

$$10x - 11(n - x) = 21x - 11n.$$

Since each individual bet carries a 4.55% HA, so too does the collection of bets. By contrast, the HA for a parlay card, assuming a 50% probability of picking each game correctly, is far higher. If you can afford the increased cost, your chance of winning money is far better by making separate wagers.

For a 10-game card, you make a profit on the set of wagers provided that you can pick 6 or more games correctly.. The expected return is

$$E = \sum_{k=0}^{10} (21k - 110) \cdot \binom{10}{k} \cdot p^k \cdot (1-p)^{10-k},$$

a 10th-degree polynomial in the probability of success p.. This polynomial is greater than 0 if $p > .5238$, so it's not necessary to be much better than a coin flip at picking winners to turn this into a positive-expectation bet.

Twenty-Six was a popular game in the Midwest (in underground gambling operations, of course) in the 1950s [103]. To play Twenty-Six, you select a number from 1 to 6—called your *point*—and roll 130 six-sided dice. In practice, this was done by having the player roll ten dice 13 times. If your point comes up 26 times or more, you are paid off at 3 to 1 odds. If you roll your point 33 or more times, you are paid 7 to 1. At the other end of the scale, if you roll your point 11 or fewer times, that pays 3 to 1 also, and if you roll your point exactly 13 times, the payoff is 2 to 1.

Suppose that you choose the point 5. Before breaking this game down, we note that the mean number of 5s you can expect to roll in 130 tosses of a fair die is $130 \cdot \dfrac{1}{6} = 21\dfrac{2}{3}$, so the number of 5s that triggers a payoff is pretty far removed—in both directions—from the expected value. With that in mind, we note that there are four possible numerical outcomes for a \$1 bet:

- Win \$1 with 13 5s.

- Win \$3, either by rolling fewer than 11 5s or between 26 and 32 of them.

- Win \$7, by rolling 33 or more 5s.

- Lose \$1, in every other outcome.

We shall treat these four cases separately. If we let X denote the number of 5s rolled in 130 trials, we have a binomial random variable with parameters $n = 130$ and $p = \dfrac{1}{6}$.

Winning \$1 is the easiest case to compute, since this involves only a single outcome.

$$P(X = 13) = \binom{130}{13} \cdot \left(\frac{1}{6}\right)^{13} \cdot \left(\frac{5}{6}\right)^{117} = .0109.$$

To win \$3, you must roll 11 or fewer 5s, or between 26 and 32 5s inclusive. Mathematically, we seek $P(X \leqslant 11) + P(26 \leqslant X \leqslant 32)$. On the TI-84+, this is easily computed with three applications of the **binomcdf** routine. $P(X \leqslant 11)$ is available with a direct application of **binomcdf** and is .0053.

To compute the second term in the sum above, we need to find $P(26 \leqslant X \leqslant 32) = P(X \leqslant 32) - P(X \leqslant 25)$, which is done with **binomcdf**(130, 1/6, 32) - **binomcdf**(130, 1/6, 25). This probability is found to be .1746. Adding this to $P(X \leqslant 11)$ gives $P(\text{Win } \$3) = .1799$.

Winning \$7 requires rolling 33 or more 5s. $P(X \geqslant 33) = 1 - P(X \leqslant 32) = 1 - .9925 = .0075$.

Losing \$1 is the outcome in all other cases. By the Complement Rule, $P(\text{Lose } \$1) = 1 - .0109 - .1799 - .0075 = .8017$.

We then have the following probability distribution for Y, the number of dollars won:

y	-1	1	3	7
$P(Y = y)$	.8017	.0109	.1799	.0075

and the expected value of a \$1 bet is then found to be -\$.1986—a 19.86% advantage for the house.

Aside from the general rule about "gambling games always favor the house," it might be reasonable to ask if there is any way to identify particularly bad bets such as this one before risking money. One tool that may be useful is the *standard deviation* of a random variable, which is denoted by the Greek letter sigma: σ.

Definition 3.3.3. The *standard deviation* σ of a random variable X with mean μ is

$$\sigma = \sqrt{\sum [x^2 \cdot P(X = x)] - \mu^2}$$

where the sum is taken over all possible values of X.

Informally, σ is a measure of how far a typical value of X lies from the mean. Computing σ using the formula above is an arithmetically intense process:

- Compute the mean of the random variable.

- For each value of x that X can attain, multiply x^2 by $P(X = x)$ and add up the products.

- Subtract the square of the mean from this sum.

- Take the square root of the difference. Notice that the use of the square root guarantees that $\sigma \geqslant 0$.

Fortunately, it is seldom necessary to perform these calculations by hand, as calculators and computer software will readily compute σ. In the special case where the random variable is binomial, we have the following simple result:

Theorem 3.3.3. *If* X *is a binomial random variable with parameters* n *and* p, *then the standard deviation of* X *is given by*

$$\sigma = \sqrt{np(1-p)} = \sqrt{npq}.$$

While the expected value of a random variable tells us about where a "typical" value of X lies, the standard deviation gives us information about the "spread" of the values of X. A larger value of σ indicates a data set where the elements farther from the mean μ, and a smaller σ corresponds to data that are clustered close to μ.

The nature of a binomial random variable X allows us to use the mean and standard deviation to derive useful information about the distribution of the data set. For example, in considering a suitably large number of tosses of two fair dice, the distribution of sums is approximately bell-shaped (many values near the mean, and fewer values as we move away from the mean of 7 in either direction) and symmetrically distributed about the mean. See Figure 3.7, which illustrates the results of 1000 simulated rolls of 2d6, for an example.

The symmetry of the graph becomes more pronounced as the number of rolls increases; Figure 3.8 shows a histogram of 10,000 rolls of 2d6, which is more symmetric than Figure 3.7.

Data sets that are distributed this way are called *normal.* In this circumstance, a result called the *Empirical Rule* is a good description of the data.

Theorem 3.3.4. *(The Empirical Rule) If the values of many samples of a random variable with mean* μ *and standard deviation* σ *are bell-shaped and symmetric, then we have the following result:*

1. *Approximately 68% of the data points lie within 1 standard deviation of the mean: between* $\mu-\sigma$ *and* $\mu+\sigma$. *(This is often stated as "approximately 2/3.")*

2. *Approximately 95% of the data points lie within 2 standard deviations of the mean: between* $\mu - 2\sigma$ *and* $\mu + 2\sigma$.

3. *Approximately 99.7% of the data points lie within three standard deviations of the mean: between* $\mu - 3\sigma$ *and* $\mu + 3\sigma$. *(This is often stated as "almost all.")*

For many practical purposes, an experimental result is deemed to be statistically significant if it is at least 2 standard deviations away from the mean, which means that its probability—under the assumption that there is no unusual effect and all of the deviation from the mean is due to random chance—is less than 5%. Since the random variable is symmetric, this 5% is evenly distributed between results greater than 2 SDs above μ and results less than 2

FIGURE 3.7: Results of 1000 rolls of 2d6.

FIGURE 3.8: Results of 10,000 rolls of 2d6.

SDs below μ. This 2 SD standard is a convention agreed upon by the statistics community; it does not fall out of any equation as a rigorously derived standard.

By using 95% as a minimum, what we are saying is that 19 out of 20 times that we identify a result as due to something other than random variation, we will be correct, and this level of confidence is acceptable in many lines of inquiry. Some fields may have more exacting standards: in experimental particle physics, for example, the standard for confirming a discovery is "5 sigma," or at least 5 standard deviations away from the expected value, corresponding to $P(\text{Chance event}) < \dfrac{1}{3,500,000}$, approximately.

Example 3.3.4. Reconsidering Twenty-Six in light of this new information, we can calculate the SD of the binomial random variable X:

$$\sigma = \sqrt{npq} = \sqrt{130 \cdot \frac{1}{6} \cdot \frac{5}{6}} \approx 4.25.$$

Figure 3.9 indicates that the number of 3s (for example; any number from 1 to 6 could be chosen) has an approximately normal distribution, and so the Empirical Rule may be used to analyze this game.

FIGURE 3.9: Number of 3s rolled in 10,000 rolls of 130d6.

The Empirical Rule then tells us the following:

$$P(\mu - \sigma < X < \mu + \sigma) = P\left(16\frac{11}{12} < X < 25\frac{5}{12}\right) \approx .68$$

$$P(\mu - 2\sigma < X < \mu + 2\sigma) = P\left(13\frac{1}{6} < X < 30\frac{1}{6}\right) \approx .95$$

$$P(\mu - 3\sigma < X < \mu + 3\sigma) = P\left(8\frac{11}{12} < X < 34\frac{5}{12}\right) \approx .997.$$

To win money, we must either roll 26 or more of our point, 11 or fewer, or exactly 13. Since 13 is an exact value and not an interval, the Empirical Rule is not necessary to assess the likelihood of that result. With $\mu = 21\frac{2}{3}$ and $\sigma = 4.25$, we can see that 11 is about 2.5 SDs below the mean, and so is outside the interval from 14 to 30 within 2 SDs of μ. The symmetry of the distribution tells us that $P(X \leqslant 13) \approx .025$, and this includes nonpaying values of 12 and 13 as well all money-winning values of 11 or less. The situation for 33 or more is the same; 33 lies about 2.67 standard deviations from the mean.

For rolling between 26 and 32 points, our chance of winning is certainly better—but "better" is always a relative term, and here it's relative to "less than a 2.5% chance." The Empirical Rule, by itself, suggests that this probability is less than 16%—hardly a winning proposition. ∎

Craps

The binomial distribution can also be useful in analyzing certain craps wagers. The probability of rolling a 7 at craps is $p = \frac{1}{6}$, and so the probability of not rolling a 7 is $q = \frac{5}{6}$. If a point has been established and you have a bet on the pass line, you are rooting for the shooter not to throw a 7.

While no one can toss a pair of fair dice under casino conditions and bring up the desired number with 100% accuracy, there are dice experts who claim the ability to control the throw of the dice so that the proportion of 7s is statistically significantly less than the 16.67% expected by chance. The stakes are not insignificant: If a shooter can cut the proportion of 7s rolled from 1 in 6 to 1 in 7, he or she can achieve an edge of 4.4% over the casino [153]. Since the number of 7s in a fixed number of tosses is a binomial random variable, we have the mathematical model to assess these shooters' skill level.

A controlled shooter's regimen has two components: *setting* the dice and *tossing* them. Setting the dice refers to how the dice are aligned in the shooter's hand, relative both to each other and to his or her fingers. In a common set intended to throw hardway combinations, the dice are held together as a rectangular solid, arranged with hardway pairs from 2-2 through 5-5 together on the outside and the 1 and 6 faces on the vertical sides. The dice, when tossed, rotate about an axis through the 1 and 6 faces, and if thrown properly, turn up more hardways and fewer 7s than expected by random chance, since the 1-6 combination is less likely [152].

A controlled toss is intended to minimize, if not totally eliminate, random bouncing of the dice. At the craps table, it is expected that the dice will rebound off the far wall of the table, which is covered with pyramid-textured rubber to induce random rolling, before coming to rest. In an ideal controlled toss, the dice leave the shooter's hand together, fly through the air touching or nearly touching with some backspin, bounce off the table and hit the wall nearly together, and finally drop back to the table as gently as possible, without much random effect imparted from the wall.

Suppose that a controlled shooter claims the ability to throw fewer 7s than the 1 in 6 that would be expected from throwing the dice at random. How would we assess their performance, ideally before risking a lot of money based on their claim? How many fewer 7s than expected would be enough to convince us that there's a real effect here, and not just random fluctuation?

In [152], Stanford Wong describes an experiment testing his and another shooter's ability to toss fewer 7s than expected by chance. In 500 rolls, the two men tossed only 74 7s. Does this constitute a statistically significant result?

The proportion of 7s in this experiment was $\dfrac{74}{500} = .1480$, certainly less than $\dfrac{1}{6}$. Since the random variable is binomial, we can use Theorems 3.3.2 and 3.3.3 to compute the mean and standard deviation.

$$\mu = np = 500 \cdot \frac{1}{6} = 83.333.$$

$$\sigma = \sqrt{npq} = \sqrt{500 \cdot \frac{1}{6} \cdot \frac{5}{6}} = 8.333.$$

It follows that the probability of tossing fewer than

$$\mu - 2\sigma = 83.333 - 2 \cdot 8.333 = 66.667$$

7's is .025, so a result of 74, while considerably less than 83.333, does not quite rise to the level of statistical significance.

In devising this challenge, Wong was aiming at a slightly lower level of significance, and set his target number at 79, roughly half a standard deviation below the expected number. The probability of tossing 79 or fewer 7s in 500 tosses is about .37—not statistically significant, but perhaps meaningful enough to give an advantage to gamblers betting on the pass line with him throwing the dice [152]. Recall that the HA for a bet on the pass line is only 1.41%, so it wouldn't take a large deviation from random chance to tip the edge over to the players.

Roulette

Controlled dice shooting is not illegal. Neither is *wheel clocking*, or repeatedly observing and recording the successive spins of a roulette wheel in an effort to detect which numbers, if any, the wheel favors through simple mechanical slackness. While professional-grade roulette wheels are very carefully machined to ensure totally random results, they are nonetheless physical objects and so subject to mechanical wear and tear that could potentially lead to some numbers appearing more often than random chance would suggest. Casino-quality roulette wheels cost thousands of dollars, and so are not replaced as often as cards and dice are.

How often that "more often" is can be determined through the use of the binomial formula together with the Empirical Rule. To do this effectively, it is necessary to consider a fairly large number of spins, since the probability of spinning a given single number is only $p = \dfrac{1}{38}$.

Example 3.3.5. If you clock a wheel through 200 spins, the mean and standard deviation of the number of times any one number appears are

$$\mu = np = 200 \cdot \frac{1}{38} = \frac{200}{38} \approx 5.263$$

and

$$\sigma = \sqrt{npq} = \sqrt{200 \cdot \frac{1}{38} \cdot \frac{37}{38}} \approx 2.264.$$

Figure 3.10, which displays the results of 10,000 trials in which a simulated roulette wheel was spun 200 times and the number of 0s recorded, shows that the distribution of the number of 0s in 200 spins is roughly normal. Accordingly, the Empirical Rule applies.

Under the assumption that the wheel is truly random, the rule can be used to show that the probability that the number of 0s is more than 3 standard deviations above the mean is approximately .0015. Accordingly, if your wheel clocking shows any number appearing more than $\mu + 3\sigma \approx 12.054$ times in 200 spins, you may safely conclude that the roulette wheel favors that number, and you should bet in line with this discovery. ∎

Nonrandomness in roulette wheels can be eliminated with an electronic game where the winning number is chosen by a random number generator.

3.4 Exercises

Answers to starred exercises begin on page 287.

FIGURE 3.10: Results of 10,000 runs of 200 spins of a roulette wheel.

3.1.* *Sicherman dice* are a variant on ordinary six-sided dice. One die of the pair is numbered 1, 2, 2, 3, 3, 4; and the other is numbered 1, 3, 4, 5, 6, 8.

a. Find the probability of rolling a 7 with a pair of Sicherman dice.

b. Find the probability of rolling doubles with a pair of Sicherman dice.

c. Let X be the sum of the numbers rolled on a pair of Sicherman dice. Construct a probability distribution for X. How does this compare to the distribution for the sum of two standard dice seen in Example 3.1.5?

3.2. Consider a game of chance where you roll a d6 once and are paid, in dollars, an amount equal to the number thrown. What would be the correct price to charge to make this a fair game?

3.3.* A one-roll craps bet that the next roll will be an 11—called "yo-leven" at the tables—pays 15 to 1. Find the house advantage.

3.4. Some older craps layouts offer space at the corners for "Big Six" and "Big Eight" multiroll bets. Big Six pays even money if a 6 is rolled before a 7, and Big Eight pays even money if an 8 is rolled before a 7. Since the 6 and 8 are equally probable, the two bets are mathematically equivalent. Of course, they are inferior to place bets on the two numbers, since the payoff odds are lower. New Jersey state law forbids the Big 6 and Big 8 bets, perhaps out of a desire to protect gamblers from a particularly disadvantageous bet.

a. Find the house advantage for the Big Six bet and confirm that it's higher than the HA for a Place bet on 6.

b. At what rate would this bet have to pay off to make it a fair bet?

3.5.* How does the HA for a $5 place bet on 9 compare to the HA for a $6 place bet on 6 calculated on page 107?

3.6.* Some casinos bar the 3 instead of the 12 on the Don't Pass bet. How does this change the probability of winning? Find the expected value of this bet variation.

3.7.* The Vermont Lottery is among many state lotteries offering a Daily 4 game. Like the Daily 3, players pick a four-digit number for one of two drawings per day. Bets on this number may be made for 50¢, $1, $2, or $5.

- For a straight bet, the player wins if their number is exactly the one matched by the state. This pays off at 5000 for 1.

- There are several varieties of boxed bets available.

  - A *24-way* boxed bet is made on a four-digit number with four different digits, such as 1729. If any combination of the four digits is chosen, the bet pays off at 208 for 1.

  - A *12-way* boxed bet covers a number which has two copies of one digit and two nonmatching digits, for example, 1146. If any of the 12 rearrangements of the number hits, the payoff is 416 for 1.

  - The *6-way* boxed bet is for a four-digit number with two pairs of identical digits, as in 6688. The payoff on a 6-way box is 834 for 1.

  - Finally, a *4-way* boxed bet can be made on a number where one of the two different digits is repeated three times, like in 3393. This payoff is 1250 for 1.

a. Compute the HA for the straight bet. How does it compare to the HA for the straight bets in the Michigan Daily 3 game (Example 3.2.18)?

b. How do the HAs for the various boxed bets compare to one another and to the HA for the straight bet on the same four-digit number?

3.8.* Figure 3.11 shows a glass panel from a roulette machine promoting its "Super 7–11" bet.

Note that this wager extends over 2 consecutive spins, provided, of course, that the first one wins.

a. Find the probability of winning this bet, assuming an American wheel.

b. What are the payoff odds on the second bet, on the number 11? How does this compare to the 35–1 payoff on the first bet?

c. Find the house advantage for this bet.

FIGURE 3.11: Super 7–11 bet from an electronic roulette machine.

3.9.* A policy game wager on two numbers instead of 3 is called a *saddle*, and pays 32 for 1 if both numbers are drawn [71]. Calculate the house advantage of a saddle bet.

3.10.* The *Capital saddle* bet is a variation on the saddle bet (Exercise 3.9) where the player's 2 numbers must appear among the first 3 numbers drawn. This bet pays 500 for 1. Find the probability of winning a Capital saddle bet.

3.11.* A 4-number policy game bet is called a *horse* bet [71]. Find the expected value of a 1-unit horse bet paying 680 for 1.

3.12. Some policy game operations drew more or fewer than 12 numbers, anywhere from 11 to 15. If a game draws n numbers, find the probability of winning a gig bet (3 numbers) as a function of n.

3.13. Just up the road from Chicago, the *Milwaukee Wheel* put a slightly different spin on the policy game [90]. In this game, the operators chose two sets of 12 numbers and a set of 6 numbers from 1–78. Players, as in Chicago, chose 3 numbers. If a gambler's numbers appeared in once of the two sets of 12, they were paid 100–1. The "cyclone" occurred when the player's numbers all fell among the set of 6, and paid 845 for 1.

a. Find the probability of a 100–1 payoff.

b. What is the chance of hitting the cyclone? (Hint: Consider the cyclone as being drawn first.)

c. Find the house edge on a Milwaukee Wheel bet.

3.14.⋆ *Fahfee* is an illegal variation on the Numbers Game that is played in South Africa. The game is quite simple: players choose a number from 1 to 36, and are paid 27–1 if their number matches the one drawn by game operators. A winning player's initial wager is returned with his payoff, so the game pays 27 to 1, not 27 for 1 [74]. Find the house advantage of a 1 rand (South Africa's currency unit) fahfee wager.

3.15.⋆ In October 2022, Big Red Keno in Beatrice, Nebraska offered a Quarter Mania promotional 5-spot keno game, which is compared to the standard 5-spot game in Beatrice in Table 3.23. Tickets for this game were priced at 25¢. Since two payoffs have increased while the third is unchanged, the house

TABLE 3.23: Beatrice, NE 5-spot keno pay table, October 2022

Standard Game		Quarter Mania	
Match	Payoff	Match	Payoff
5	$200	5	$250
4	$1.00	4	$1.75
3	$0.25	3	$0.25

advantage has gone down. How much has the house advantage decreased with the promotional pay table?

3.16.⋆ The Wyoming state lottery introduced keno in 2022. Its 5-spot pay table is shown in Table 3.24. How does the house edge for this game compare

TABLE 3.24: Wyoming Lottery 5-spot keno payoffs, $1 wager

Match	Payoff
5	$500
4	$16
3	$2

to the standard HA in Beatrice, NE (Exercise 3.15)?

3.17.⋆ In Table 3.15, explain why switching the "Full house" and "4-card royal flush" lines would not change the strategy for a player following the table.

3.18. A parlay card offered by a consortium of seven casinos in southern Nevada offers the payoffs indicated in Table 3.25.

Let p be your probability of picking the winner of a single game.

a. What value of p will make the 3 for 3 bet a break-even proposition?

b. What value of p will make the 9 for 9 bet a break-even proposition? If you have that same probability p of picking one game correctly, do you have an advantage in playing the 10 for 10 wager? If so, how big is it?

TABLE 3.25: Alternate sports parlay card payoffs

Wins	Payoff
3 for 3	6 for 1
4 for 4	10 for 1
5 for 5	20 for 1
6 for 6	35 for 1
7 for 7	75 for 1
8 for 8	125 for 1
9 for 9	250 for 1
10 for 10	700 for 1

3.19.* Compute the house advantages for the bets other than on the $1 spot on the Big Six Wheel, using the Las Vegas payoffs stated on page 102.

3.20.* For an n-team teaser card paying off at a to b, derive an expression in terms of a, b, and n for the minimum probability p of success at picking a winning team with the additional points that gives the wager a positive expectation.

3.21. In 1994, two Las Vegas casinos run by the Hilton Corporation charged over/under sports bettors 12 to win 10 rather than the usual 11 on basketball games, citing a belief that over/under lines could be exploited by skilled sports bettors with access to game information such as injuries to key players [67]. Show that doubling the excess charge from $1 to $2 increases the 4.55% house advantage, but does not double it to 9.10%.

3.22. Example 2.4.13 described the Ninety Percenter punchboard. The punchboard has 800 spots, which are sold for 10¢ each. Players punching out any multiple of 10 between 10 and 650 receive a $1 prize, and the last spot on the board carries a $5 bonus.

Assuming that the advertised 90% return is accurate, calculate the expectation of the first bet on a fresh Ninety Percenter punchboard.

3.23.* For the Ninety Percenter punchboard described in Exercise 3.22, calculate the expectation for the player punching out the second spot.

3.24. Recall that the probability of drawing a royal flush at video poker is approximately 1 in 40,000. Use this estimate to calculate the probability of getting exactly two royal flushes in 40,000 hands of video poker.

3.25.* In Example 2.4.3, we considered the parlay card available in the sports books run by Station Casinos. Assuming a 50% chance of picking each game successfully, find the probability that you will pick exactly half of a ten-game parlay correctly.

3.26. Another way of setting two dice is designed to throw more 7s than the 1 in 6 fraction we expect from a random roll. This could be an advantage for the shooter on the come-out roll. In 400 rolls, how many 7s would be necessary to constitute a statistically significant excess?

3.27. In spider craps (see page 30), the payoff for a one-roll bet that the next roll will be a 2 is 55 to 1. Find the house advantage for this bet. How does the HA compare to the 13.89% HA for the corresponding wager at standard craps?

3.28. The spider craps equivalent of the craps field bet pays off if the next roll is 2–6 or 12–16, with a 2 to 1 payoff on a 16 and an even-money payoff on the other rolls. Find the house edge on this wager.

3.29. The one-roll *horn* bet in spider craps is derived from the standard craps menu of wagers. In standard craps, a horn bet wins if the next roll is a 2, 3, 11, or 12; in spider craps, the horn bet covers the numbers 2, 3, 15, and 16. Customarily, horn bets are made in multiples of 4 units, with equal bets on each number. At least three of these four bets will lose on each roll, but the fourth may win and cover the other losses. Bets on 2 or 16 pay off at 55 to 1; bets on 3 or 15 pay off at 28 to 1. Find the HA of a $4 horn bet at spider craps.

3.30. At what odds must the baccarat Tie bet pay off to reduce the HA on this wager to approximately 4%?

3.31. Suppose that a variation on baccarat provides that bets on Player or Banker both lose rather than push if the two hands tie. Find the HA for both bets with this rule in force.

Street Craps

Casino craps shares many features with the street game of craps, which is considerably less formal and offers fewer protections to gamblers. Street craps games are often patronized by craps hustlers, who offer side bets to players that frequently appear much fairer than they are.

3.32. One such proposition, described in [111], is made when a point of 6 or 8 is established. The hustler offers even money that the point will not be made in 3 rolls or fewer. A player accepting this bet backs the shooter to roll the point before a 7 within 3 rolls; if the bet is not resolved in 3 rolls, the bet pushes.

a. Find the probability that a shooter fails to make a point of 6 or 8 in 3 or fewer rolls.

b. Find the probability that a shooter makes a point of 6 or 8 in 3 or fewer rolls.

c. Calculate the hustler's edge on this bet.

3.33. Repeat Exercise 3.32 for a point of 5 or 9 in 4 rolls.

3.34.* Consider the proposition that a player will roll a 7 or 8 in 2 rolls. This bet can be made independent of whether or not a point is active. A dice hustler offers to take the "will roll" side and offers even money to players wishing to bet that the shooter will not roll a 7 or 8 in 2 rolls. Find the hustler's advantage.

3.35.* A hustler may offer 3–2 that a point of 4 or 10 will not be made [111]. What is the expectation for the player who accepts this proposition and bets that the point will be made?

Chapter 4

Modified Casino Games

In the preceding chapters, we considered what might be called "standard" casino games. Game development specialists frequently propose new games or modifications to existing games or wagers, which are the focus of this chapter.

As of August 2021, there are over 1250 table games and game variations approved for play in Nevada casinos [83]. While the vast majority of new casino game ideas are unsuccessful, the potential to make a lot of money by selling the rights to a new popular game continues to attract game designers. For example, the inventors of Caribbean Stud Poker sold their idea for $30 million [57]. An important part of any new game proposal is a mathematical analysis of the bets and a careful calculation of the HA.

4.1 Wheel Games

Roulette

It is common practice in Oklahoma tribal casinos to charge a per-hand fee at card games independent of a player's wager, and this practice was carried over to craps and roulette tables when those games were permitted beginning in 2018 [15]. In roulette, this takes the form of a $1 ante charged per spin of the wheel. Combined with a $5 minimum bet, the effective payout on a winning even-money bet was 5–6. The expected value of such a bet was then

$$E = (5) \cdot \frac{18}{38} + (-5) \cdot \frac{20}{38} - 1 = -\frac{48}{38},$$

giving a house advantage of $\frac{48}{190}$, or 25.26%—nearly 5 times the 5.26% advantage without the fee.

Since the fee was fixed rather than set as a proportion of a player's wager, it was possible to decrease the casino's advantage by making a bigger bet. On an N even-money bet, the expectation becomes

$$E(N) = (N) \cdot \frac{18}{38} + (-N) \cdot \frac{20}{38} - 1 = \frac{-2N - 38}{38}.$$

DOI: 10.1201/9781003358183-4

Dividing by N gives a HA of

$$-\frac{-2N-38}{38N} = \frac{1}{19} + \frac{1}{N},$$

which decreases as N increases. For a \$100 wager, the player is fighting a house edge of 6.26%.

How does the ante affect other roulette bets? A \$5 single-number bet would have an expected value of

$$E = (175) \cdot \frac{1}{38} + (-5) \cdot \frac{37}{38} - 1 = -\frac{48}{38},$$

as with even-money bets. The HA remains 25.26%, and it decreases in inverse proportion to the amount of the wager as well.

The same mathematics applies to every bet on an American wheel, except for the basket bet. The expected value of a \$5 basket bet in Oklahoma is

$$E = (30) \cdot \frac{5}{38} + (-5) \cdot \frac{33}{38} - 1 = -\$\frac{53}{38} \approx -\$1.39,$$

leading to a house advantage of 27.89%.

A new roulette betting option, introduced in 2008 and seen at the Orleans Casino in Las Vegas, is the "Colors" bet [12]. To make this bet, a gambler wagers on either Red or Black, and if that color then turns up on three consecutive spins, the payoff is 8 to 1. We would like to find the house advantage for a \$1 Colors bet if an American (0 and 00) roulette wheel is used.

No matter which color is chosen, we have $P(\text{Win}) = p = \frac{18}{38}$ for a single spin. Since the spins are independent, the probability of winning the Colors bet is $p^3 = \left(\frac{18}{38}\right)^3 = \frac{729}{6859} \approx .1063$, and so we have

$$E = (8) \cdot \frac{729}{6859} + (-1) \cdot \frac{6130}{6859} = -\frac{298}{6859} \approx -\$.0434,$$

and so the house advantage is about 4.34%—somewhat less than the HA for any other bet on the board.

As with other roulette wagers, the HA depends on the wheel. An analogous calculation for European roulette wheels, where $P(\text{Win})$ ticks up to $\frac{18}{37}$, shows that the expected value of a \$1 Colors bet is

$$E = (8) \cdot \frac{5832}{50,653} + (-1) \cdot \frac{44,821}{50,653} = \frac{1835}{50,653} \approx \$.0362,$$

about 3.6¢. Since this is positive, the player has a 3.6% advantage over the casino—meaning that you won't see the Colors bet with this payoff on a European wheel. If you do, someone has made a serious calculation or judgment error. The bet is unfair in your favor, and you should settle in and prepare to win money.

A challenge facing the casino game designer who wishes to develop a variation on roulette is the increased cost of a nonstandard wheel. Professional-grade roulette wheels in European or American configurations cost thousands of dollars; there would be considerable additional expense in manufacturing a specialized wheel with a different number of pockets. While a large number of alternate casino games have been marketed using standard cards and dice in new ways, new roulette games—as opposed to new wagers using a standard wheel—often require a modified wheel, and so are considerably rarer.

Royal Roulette

Royal Roulette is a variation on roulette, developed in Australia, that uses familiar playing cards as spots on the wheel [99]. The wheel contains 50 pockets: one for each of the 48 cards from 2 through king, a single ace, and a joker. The layout (Figure 4.1) is similar to a traditional roulette layout and offers the betting options listed in Table 4.1.

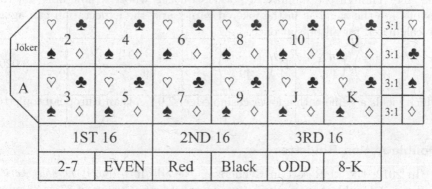

FIGURE 4.1: Royal Roulette layout

How does Royal Roulette compare to the standard game on which it is based? A quick examination of the payoff structure shows that the payoff on a winning bet on n cards is $\dfrac{48-n}{n}$ to 1. There is no exception corresponding to the basket bet in American roulette. Royal Roulette's versions of the basket bet, by which we mean unusual bets derived from the layout, are the Court and Six Line bets. Unlike in standard American roulette, the Royal Roulette analogs pay in line with other bets. We can compute the expected value of any Royal Roulette bet with one simple equation:

$$E = \left(\frac{48-n}{n}\right) \cdot \left(\frac{n}{50}\right) + (-1) \cdot \left(\frac{50-n}{50}\right) = \frac{48-n-50+n}{50} = -\frac{2}{50} = -\$.04$$

—for a house edge of 4%, a result which is independent of n and more favorable to the player than American roulette.

TABLE 4.1: Royal Roulette betting options

Name	Payoff	Number of cards covered
Straight Up	47 to 1	1, including Ace or Joker
Split	23 to 1	2
Court	15 to 1	3: Ace 3♠ 3♡ or Joker 2♠ 2♡
Street/Corner	11 to 1	4
Six Line	7 to 1	6: Ace, Joker, 2♡, 2♠, 3♡, 3♠
Eight Line	5 to 1	8
Suit/Column	3 to 1	12: One suit or one column
Sixteen Set	2 to 1	16
Even Chances	1 to 1	24: Red, Black, Odd, Even, 2-7, 8-K

A second version of Royal Roulette removes the joker from the wheel and the layout—we might think of this, informally, as European Royal Roulette. The Six Line bet is eliminated, though the Court bet remains and must involve the Ace. Though the number of pockets on the wheel is now only 49, the payoffs are the same as in the jokered game, and the expectation can again be calculated for all bets at once:

$$E = \left(\frac{48-n}{n}\right) \cdot \left(\frac{n}{49}\right) + (-1) \cdot \left(\frac{49-n}{49}\right) = \frac{48-n-49+n}{49} = -\frac{1}{49} \approx -\$.0204,$$

which yields a game with a lower casino edge, 2.04%, than European roulette.

Double Action Roulette

In 2012, the M Resort in Henderson, Nevada introduced *Double Action* roulette. A Double Action wheel contains two identical rings of 37 or 38 numbers (in either European or American roulette configurations). The two wheels rotate independently in opposite directions, and when the ball falls into a pocket, it identifies two winning numbers, one on each wheel.

Two separate betting layouts allow standard roulette bets (with an HA of 2.70% or 5.26%) on either wheel separately, and a collection of new bets that apply to the numbers on both wheels. These bets mimic the standard roulette bets; one may bet, for example, on whether or not both numbers will be red, or even, or low.

One new bet is a separate single-number *side bet* (a bet unrelated to the main play of a casino game) that pays off at 1200 to 1 when the chosen number is hit on both wheels in a single spin. The expectation on a $1 bet on this proposition, at an American Double Action wheel, is

$$E = (1200) \cdot \left(\frac{1}{38}\right)^2 + (-1) \cdot \left[1 - \left(\frac{1}{38}\right)^2\right] = -\frac{243}{1444} \approx -\$.1683,$$

a 16.83% house edge. The edge for this bet on a European Double Action wheel is not much better: 12.3%. In the first month of Double Action Roulette's test run at M, there were 18 of these jackpots paid [110].

Double Action is a direct descendant of *Double Roulette*, a game introduced at Monte Carlo in 1936, which used two concentric wheels and two roulette balls [22]. The high house advantage doubtlessly contributed to the game's short lifespan in Monaco.

Roulette 73

Roulette 73, as the name suggests, offers a wheel with 73 pockets, numbered 1–72 and 0. The layout is a 6 by 12 grid topped with a rectangle for bets on 0 (Figure 4.2), so there are 6 possible column bets and street bets include 6 numbers.

0					
1	2	3	4	5	6
7	8	9	10	11	12
13	14	15	16	17	18
19	20	21	22	23	24
25	26	27	28	29	30
31	32	33	34	35	36
37	38	39	40	41	42
43	44	45	46	47	48
49	50	51	52	53	54
55	56	57	58	59	60
61	62	63	64	65	66
67	68	69	70	71	72

FIGURE 4.2: Roulette 73 layout.

Three-number bets are only available by combining 0 with two consecutive numbers from 1 to 6. As in American and European roulette, a wager on n numbers has a payoff in inverse proportion to the area covered, in this case $\frac{72-n}{n}$ to 1. The common HA of these wagers is then

$$E = \left(\frac{72-n}{n}\right) \cdot \frac{n}{73} + (-1) \cdot \frac{73-n}{73} = -\frac{1}{73} \approx -.0137,$$

giving a 1.37% HA that outstrips the standard HA on any roulette wheel we've seen so far.

As shown in Figure 4.2, 24 of the numbers are colored gray on the Roulette 73 layout, though they are red or black on the wheel. These 24 numbers are the basis for a 14-chip combination bet. This bet includes 4 single-number bets: on the numbers 8, 13, 44, and 49, together with 10 chips in split bets covering the other 20 numbers in pairs. Since expected value is additive (Theorem 3.2.1), we can compute the expected value of this combined bet by looking at the sum of the expected values of its components.

With 14 bets in play, each with the same expected value, the expectation with \$1 chips is $E = 14 \cdot \left(-\dfrac{1}{73} \right) = -\dfrac{14}{73} \approx -\$0.1918$, but when we divide by the \$14 wagered, we find ourselves looking again at a bet with a 1.37% HA.

Big Six Wheel

The symbol distribution and corresponding payoff values on a Big Six Wheel are certainly open for modification as a casino may wish. In particular, there is no requirement that payoffs correspond to currency denominations, or that currency symbols be used at all.

One possible reason for the high HA on the classic Big Six Wheel is that the game attracted less player traffic than other table games, yet incurred the expense of a full-time casino employee who spent considerable time alone, waiting for gamblers to try their luck. With the advent of all-electronic games such as Big Six Wheels manufactured by Interblock, newer wheels require no human dealer. The cost of operating the game is then decreased considerably, and some of the savings can be passed on to patrons in the form of better payoffs.

A wheel at the Soaring Eagle Casino used the symbols shown in Table 4.2. Numbered spaces pay their numerical value; the American flag and Joker spaces pay 50–1.

TABLE 4.2: Big Six wheel symbols: Soaring Eagle Casino

Symbol	Count
1	26
3	13
6	7
12	4
25	2
Flag	1
Joker	1

One can immediately see that some payoffs are greater than currency values, which might suggest a more player-friendly game. This wheel is somewhat better for players than the standard Big Six wheel; a bet on the \$1, \$3, \$12, or \$25 spots faces a HA of only 3.70%; the house edges on the other bets are all less than the 11.11% HA of the best standard Big Six bet. A second advantage to a better pay table is that patrons might win more frequently or win larger amounts. Passersby might then be drawn in by the attendant excitement, and some might be encouraged to play a game that has a reputation for being unfriendly to players.

Dream Catcher, from Evolution Gaming, is a Big Six wheel variation in a live online casino, where a dealer spins the wheel and interacts with online players. The wheel includes two bonus multiplier sectors, which replace the special logos. The breakdown of its prizes is shown in Table 4.3.

TABLE 4.3: Dream Catcher wheel symbols [72]

Symbol	Count
1	23
2	15
5	7
10	4
20	2
2×	1
7×	1

If a wheel spin turns up either one of the multipliers, all bets are held and the wheel is respun until a number appears—subsequent spins landing on multipliers are disregarded. Once a winning number is chosen, all payoffs are multiplied by that multiplier, so a \$1 bet on the 20 spot could pay off \$140 if the 7× and 20 sectors are spun in succession. The probability of this payoff is

$$\frac{1}{54} \cdot \frac{2}{54} = \frac{1}{1458}.$$

Unlikely though this may be, the presence of the multipliers nonetheless decreases the large HA on the wheel.

Example 4.1.1. Recall that the standard Big Six configuration has an 11.11% HA when betting on the 1. Since successive spins of the Dream Catcher wheel are independent, the probability of winning \$7 when betting on the 1 spot is

$$\frac{1}{54} \cdot \frac{23}{52} = \frac{23}{2808}.$$

Here, the denominator in the second fraction, which represents the chance of spinning a 1 after the 7× sector has come up, is 52 rather than 54 since the multiplier spaces are disregarded on that spin. The complete probability distribution for this bet is shown in Table 4.4.

The HA of this bet is then only 5.80%, about half of the HA on a standard wheel. ∎

Another alternate electronic wheel uses the wheel and clapper only as an indicator of the winning result chosen by a random number generator. A wheel

TABLE 4.4: Dream Catcher probability distribution: $1 bet on the 1 spot

Outcome	Probability
$1	$23/54$
$2	$23/2808$
$7	$23/2808$
-$1	$29/52$

at Circus Circus in Las Vegas in 2022 has 54 sectors, distributed as per Table 4.5.

TABLE 4.5: Alternate Big Six Wheel configuration.

Symbol	Count
1	24
3	12
5	8
11	4
23	2
Silver diamond	1
Red diamond	1
Superspin	2

The two diamond spaces pay 45–1. The Superspin spaces offer the possibility of a large payoff, though the probability of hitting a Superspin space is only $\frac{2}{54}$. If a Superspin sector is selected, the wheel activates a second bonus spin where bettors who made a Superspin bet can win 20, 50, 75, 100, 500, or 1000 times their initial wager. The PDF for this second spin is a trade secret; one might reasonably assume that the 6 possible outcomes are not equally likely. Nonetheless, this opens up the possibility of a much bigger player win than a traditional wheel offers.

4.2   Dice Games

Crapless Craps

Crapless craps is a variation that was described in a 1930 work by Charles E. Shampaign, where it was called "Everything A Point" [122]. The game was refined and brought to the casino floor by Bob Stupak, owner of the Vegas World Casino (now the Strat) in Las Vegas. If 2, 3, 11, or 12 is rolled on the come-out roll, they neither lose nor win, but become points like any other point, and the shooter must roll that number again before rolling a 7 to win. A 7 on the come-out roll remains an automatic winner. To make up for this change in rules, the Don't Pass bet is not offered at crapless craps—you cannot bet against the shooter.

Example 4.2.1. On the face of it, this looks like a good deal for dice players: four automatic losing come-out rolls (1-1, 1-2, 2-1, and 6-6) are now points with a chance to win, while only two automatic wins (5-6 and 6-5) have been converted to possible losers. The mathematics tells a different story, though. In addition to the figures in Example 1.5.2, we have the following new probabilities:

Point	P(Point rolled)	P(Point wins)	P(Win on this point)
2 or 12	$1/36$	$1/7$	$1/252$
3 or 11	$2/36$	$2/8$	$1/72$

The probability of winning after a come-out roll of 2 or 12 has risen from 0 to $\frac{1}{7}$, and the probability of winning after a 3 has risen from 0 to $\frac{1}{4}$. However, the probability of winning after an 11 has fallen from 1 to $\frac{1}{4}$. The player has given up far more than he has gained. Adding everything up reveals that the probability of winning a Pass line bet at crapless craps is .4730, and thus that the house advantage has risen to approximately 5.39%—making this game about as advantageous for the casino as American roulette. ∎

Free Odds

Free odds bets in craps, though commonly available, are a casino rarity—a bet with zero house advantage. Once a point is established on the come-out roll, players who have placed a pass line bet have the opportunity to "back it up" with an additional odds bet. These odds bets are limited to a certain multiple of the original bet, often "3X/4X/5X," meaning three times the original bet if the point is 4 or 10, four times the original bet if the point is 5 or 9, and five times the original bet if the point is 6 or 8.

Putting the "free" in free odds is this: These bets are fair bets, without any house advantage. An odds bet on a point of 4 or 10 pays 2 to 1, since the probability of making a 4 or 10 is $\frac{3}{9} = \frac{1}{3}$. Converting to odds, the odds against making a 4 or 10 are 2 to 1—precisely the payoff odds. Similarly, an odds bet on 5 or 9 pays off at the correct odds of 3 to 2, and odds bets on 6 or 8 are paid at 6 to 5. It is worth noting that, here as elsewhere in a casino, fractions of a dollar that cannot be accommodated by common chip values—casino chips with values less than 50¢ are unusual; some casinos use half-dollar coins for this denomination—are rounded *down*, in favor of the casino. To gain the full effect of free odds, it is important to bet in such a way that no rounding is necessary. The easiest way to do this is to remember to place odds bets in multiples of $10.

Free odds are also available on Don't Pass bets; since the Don't Pass bettor has the advantage once the point is established, these pay off at less than even money: 1 to 2 if the point is 4 or 10, 2 to 3 on 5 or 9, and 5 to 6 on 6 or 8. Moreover, odds bets backing up a Don't Pass bet may be taken down by the bettor at any time before they're resolved—casinos are pleased to let gamblers back out of a bet where they have a greater chance of winning than of losing.

The 3X/4X/5X structure makes paying off winning pass line bets with full odds easy for craps dealers. If the initial bet is made for 1 unit and backed up with full odds, Table 4.6 shows that the total payoff is always 7 units [109].

TABLE 4.6: Craps payoffs with a 1-unit bet and full odds

Point	Max odds bet	Payoff odds	Total payoff (units)
4, 10	3X	2–1	7
5, 9	4X	3–2	7
6, 8	5X	6–5	7

Example 4.2.2. From a gambler's perspective, then, pass line bets should be made to get as little as possible on the front line and as much as possible in an odds bet. If the casino has a $5 minimum on the pass line and you wish to bet $25 on each decision, the best possible play is to make a $5 line bet and, once the point is established, back it up with a $20 odds bet, provided that the multipliers allow that. Recall that the HA on a pass line bet is 1.41%. Using free odds in this manner, and assuming that the casino offers at least 4X odds on every point (for simplicity), the expected return on the bet is

$$E = 5 \cdot \frac{8}{36} - 5 \cdot \frac{4}{36} + 45 \cdot \frac{2}{36} + 35 \cdot \frac{4}{45} + 29 \cdot \frac{50}{396} - 25 \cdot \frac{196}{495} = -\frac{7}{99},$$

which, on dividing by the $25 wagered, corresponds to an HA of merely .28%. The first two terms in the middle of this equation correspond to those outcomes when the bet is resolved on the come-out roll and there is no opportunity to place an odds bet; the remaining terms represent the different ways that the bet with full odds can play out.

By contrast, if you make a $15 pass line bet and then back it up with a $10 odds bet (to avoid fractional payoffs), the expected return is

$$E = 15 \cdot \frac{8}{36} - 15 \cdot \frac{4}{36} + 35 \cdot \frac{2}{36} + 30 \cdot \frac{4}{45} + 27 \cdot \frac{50}{396} - 25 \cdot \frac{196}{495} = -\frac{7}{33},$$

and the HA rises to .84%.   ∎

Why would a casino offer odds bets? One reason is to encourage activity at the craps tables, of course. Another is to encourage larger bets and take advantage of volatility. We note that the only time that an odds bet can be made is *after* a point has been established, which is when the casino has the advantage over a pass line bettor.

The exact multiple allowed is decided by the individual casino; in 2012, as part of a new focus on player-friendly gambling in their advertising, the Riviera Casino in Las Vegas offered 1000X odds bets on any point in an effort to entice craps players to their casino. A pass line player taking full advantage of the 1000X bets at a $5 minimum table at the Riviera was risking $5005 on each established point. Some of these bets will be lost to the casino, even though the expectation remains

$$E = (5) \cdot \frac{8}{36} + (10,005) \cdot \frac{2}{36} + (7505) \cdot \frac{4}{45} + (6005) \cdot \frac{50}{396}$$
$$+ (-5) \cdot \frac{4}{36} + (-5005) \cdot \frac{196}{495}$$
$$= -\frac{7}{99}.$$

Neither the odds of winning nor the expected value has changed, since the free odds bet is paid at true odds. Only the magnitude of the bets involved is different. The HA has dropped to a mere .0014%, but every losing odds bet nets the casino $5005, and it is quite likely that many dice players will not be able to withstand too many consecutive losses at these stakes.

The mean and standard deviation can be used to assess the volatility of this betting option. Consider the case where the point is 6 (or 8). For the wagers in which a full odds bet is made—that is, where the bet is not resolved on the come-out roll—we have the following probability distribution for the winnings X:

x	6005	−5005
$P(X = x)$	5/11	6/11

This distribution gives $\mu = -\$.45$ and $\sigma = \$5482.20$. Note the extremely high value of σ relative to μ—this indicates a bet with very high variability and thus the potential for wild swings back and forth. Since the casino has far more money than any player, it is much better equipped to withstand a run of bad luck than the gambler is.

Casino officials who are tasked with rewarding gamblers with complimentary meals, hotel rooms, and other amenities ("comps") face a tricky dilemma when confronted with evaluating the play of gamblers making pass line bets backed up by immense odds bets. Comps are typically awarded based on a percentage of the player's theoretical, not actual, losses. A gambler betting $10 on the pass line and backing it up with 100X odds of $1000 is giving the casino an average win of only .014%, even as she loses $1010 on every losing bet. Her comps will be based on the casino's theoretical return, which is 14¢ per wager, not on the amount lost [109].

Craps Side Bets

It may be said that *every* bet on a craps table except for the Pass and Don't Pass wagers is a side bet, in the sense that they are bets that are not related to the main play of the game. The bets described in this section are relatively new bets that do not have dedicated spaces on the standard craps layout (Figure 1.5).

The *Fire Bet* is a craps side bet, developed by Las Vegas casino supervisor Perry Staci, whose name suggests its purpose: to cash in on a shooter having a "hot" hand and making many points before sevening out [11]. A bet on the Fire Bet pays off if the shooter makes a certain number of *different* points, at least three or four depending on the pay table, before tossing a 7. While the pay table for the Fire Bet varies from casino to casino, one version is this:

Points Made	Payoff
4	10 to 1
5	200 to 1
6	2000 to 1

The points made must be different: if the shooter makes the points 4, 9, 4, and 6 before sevening out, this is only three different points for the purpose of the Fire Bet. A roll of 7 on a come-out roll does not interrupt a string of points. Mathematically, the Fire Bet is somewhat complicated to analyze, due to the different probabilities for establishing and making the six possible points. A good source for the computations required is [116]; there we find the following:

Points Made	Probability
0	.5939
1	.2608
2	.1013
3	.0334
4	.0088
5	.0016
6	.000162

With the pay table above, the probability of winning anything with the Fire Bet is a mere .0106, or 1.06%. The corresponding expectation is

$$E = (-1) \cdot .9894 + (10) \cdot .0088 + (200) \cdot .0016 + (2000) \cdot .000162 = -.2486,$$

giving an HA of 24.9%.

In part, no doubt, because of the high casino advantage, the Fire Bet has gained a place on the craps table in many casinos. Despite the high HA, gamblers seem to be drawn to it; they are perhaps attracted in part by the large payoff for a small bet. The maximum Fire Bet wager at most casinos offering this wager is $5.

In 1997, Donald Catlin and Leonard Frome proposed a new craps wager called the *Hard Hardway* bet. This is a simple wager that each of the six even sums, 2 through 12, will be rolled the hard way, as doubles, twice before a 7 is rolled. As one might expect, this is an extremely unlikely event. Considerable calculation (see [14]) leads to a probability of winning of

$$p = \frac{9,198,254,528,424}{1,458,015,678,282,240,000},$$

which is approximately 1/158,510.

Catlin and Frome noted that an easier route to a good estimate of the probability of winning a Hard Hardway bet was a computer simulation of the wager; they indicated that 1 billion trials would produce a sufficiently accurate approximation. In an excursion into experimental probability, the Python program in Figure 4.3 executes that simulation.

```python
import random

def dieroll():
    return (random.randint(1,6))

wins = 0
for i in range (0,1000000000): # 1 billion trials
    A = [0]*6
    crapsroll = 0
    amin = 0
    while(crapsroll != 7 and amin < 2):
        first=dieroll()
        second=dieroll()
        crapsroll = first + second
        if first == second:
            A[first-1] += 1
        amin = min(A)
    if (amin == 2):
        wins = wins + 1
        print (A[0],A[1],A[2],A[3],A[4],A[5],i)
    if (i % 100000000 == 0):
        print ('trials = ', i)
print ('Total Wins = ', wins)
print ('Probability of winning = ', wins/(i+1))
```

FIGURE 4.3: Python program for 1 billion trials of Hard Hardway [10].

This program takes about 6 hours to run. When it was executed, the first success occurred on trial number 153,637—just within the interval from trial #1 to trial #158,510 where we'd expect, on average, to find it. In this trial, there were 4 rolls of ⚁ ⚁ and 2 rolls of each other hardway total. The last win was on trial 999,707,594. In 1 billion trials, there were 6402 wins, for a win probability of

$$6.402 \times 10^{-6} \approx \frac{1}{156,201},$$

a figure within 1.48% of Catlin and Frome's theoretical value.

Either way, this seems so unlikely that no one would ever make this bet. On the other hand, the probability of winning a Powerball lottery jackpot is considerably less, and people still buy lottery tickets, so it's possible that, as with the Fire Bet, a large potential payoff might attract bettors in spite of the unfavorable odds. Suppose that a casino offers the Hard Hardway bet with a payoff of 150,000 to 1 for convenience—theirs. The expected value for a $1 Hard Hardway bet with this payoff is

$$E = (150,000) \cdot p + (-1) \cdot (1 - p) \approx -\$.054,$$

where p is the theoretical value above, and so the house advantage is about 5.4%. While this is more unfavorable to a player than a Pass or Don't Pass bet, the house advantage here is considerably less than that of several craps bets that are currently offered.

Beyond Craps

Here's a problem posed by noted gambling mathematician Peter Griffin in the magazine *Casino & Sports* and reprinted in [41]:

> You have your choice of betting that a pair of sixes will appear on a roll of 2d6 before back-to-back totals of seven appear, or taking the opposite position: that two consecutive sums of seven will be rolled before a 12. Which side should you take?

At first glance, this looks like an even proposition: the probability of rolling a 7 is $\frac{1}{6}$, and the probability of rolling two 7s in succession is thus $\frac{1}{36}$—the same as the probability of rolling a 12 on 2d6. Griffin reported that "an enterprising pair from New York was offering this proposition in Atlantic City," but that they were backing the less favorable outcome. This suggests that the proposed game is not as even as it may first appear. Which is the favored side?

Successive rolls of the dice are independent, and so the Multiplication Rule is useful in solving the problem. Denote by A the event that a 12 is rolled before two consecutive 7s. If $P(A) > .5$, this side has the advantage; if $P(A) < .5$, the other side is favored. A is itself the union of infinitely many mutually exclusive events:

- $A_1 = $ A 12 is rolled on the first throw. $P(A_1) = \frac{1}{36}$.

- $A_2 = $ A 12 is not rolled on the first throw but is then rolled on the second throw. For convenience, we divide A_2 into two subcases: the case where a 7 is rolled on the first toss and then a 12 comes up on the second, and the case where the first roll is something other than a 7 or 12 (a "neutral" roll) and a 12 is rolled on the second toss. It then follows that

$$P(A_2) = \frac{1}{6} \cdot \frac{1}{36} + \frac{29}{36} \cdot \frac{1}{36} = \frac{35}{1296}.$$

- A_3 = The first two tosses produce neither a 12 nor two straight 7s, and a 12 is then thrown on the third roll. We have

$$P(A_3) = \left[\left(\frac{35}{36} \right)^2 - \frac{1}{36} \right] \cdot \frac{1}{36} = \frac{1189}{46,656}.$$

- And so on. The probabilities become smaller as n, the number of rolls, increases, but $P(A_n)$ is never 0.

Let $P(A)$ be denoted by p. Since the events A_n are mutually exclusive, the Addition Rule gives the infinite series

$$p = \sum_{n=1}^{\infty} P(A_n) = \left(\frac{1}{36} \right) + \left(\frac{1}{6} \cdot \frac{1}{36} + \frac{29}{36} \cdot \frac{1}{36} \right) + \left(\frac{35^2 - 36}{36^2} \cdot \frac{1}{36} \right) + \cdots$$

Some algebraic gymnastics will allow us to convert this infinite sum of probabilities into something easily managed. If the first roll is a 12, the game is over, and similarly if the first roll is a 7 but the second is a 12. If the first toss is neutral (probability $\frac{29}{36}$), then we can think of the game as starting on the second roll, with $P(A) = p$. Finally, if the first toss is a 7 but the second is neutral, we're back at the beginning again, and the probability p reappears as a factor. Taken together, we have

$$p = \left(\frac{1}{36} \right) + \left(\frac{1}{6} \cdot \frac{1}{36} \right) + \left(\frac{29}{36} \cdot p \right) + \left(\frac{1}{6} \cdot \frac{29}{36} \cdot p \right).$$

This finite sum is a linear equation in p and hence is easily solved. Collecting terms on the right gives

$$p = \frac{7}{216} + \frac{203}{216} \cdot p,$$

from which we find that

$$p = \frac{7}{13} \approx .538,$$

and so it is more likely that a 12 will be rolled before a pair of consecutive 7s.

Why is 12 favored over a pair of 7s? It's because any roll has the possibility of turning up a 12, while a 7 can only win if the previous roll was a 7. For example, if the first roll was ⚁ ⚅, the probability of winning the pair of 7s bet on the next roll is 0, while the chance of winning the 12 side of the wager remains $\frac{1}{36}$.

Barbooth

Bob Stupak was known for offering unusual games at Vegas World. In the game of *barbooth*, he took a street game called *barbudi*, a Canadian game also favored in large northern U.S. cities, and revamped it for casino play. Barbooth is unusual in that it is a fair game, with neither side having an advantage [103].

In barbooth, two players alternate turns at throwing two dice. A shooter, and everyone betting on him or her, wins when tossing 3-3, 5-5, 6-6, or 5-6; the probability of this is $\frac{5}{36}$. The shooter and any backers lose when 1-1, 1-2, 2-2, or 4-4 is rolled; an event that also has probability $\frac{5}{36}$—making the game even. All other rolls result in the dice being passed to the other player. A winning player in one round shoots first in the next round. In the casino, these two players are designated "Player" and "Bank," as is also the case in baccarat.

Example 4.2.3. How many rolls, on average, are necessary for a round of barbooth to be resolved?

Define $p = \frac{10}{36}$, the probability of the game ending on any one turn, and let

$$q = 1 - p = \frac{26}{36}.$$

The probability of the game ending on the first roll is p. The probability of the game lasting two rolls is $q \cdot p$, where the first roll neither wins nor loses and the second roll finishes the game one way or the other. In general, a game has probability $q^{k-1} \cdot p$ of lasting exactly k rolls.

The expected number of rolls is then

$$E = \sum_{k=1}^{\infty} k \cdot q^{k-1} \cdot p$$

$$= \frac{p}{q} \cdot \sum_{k=1}^{\infty} k \cdot q^k$$

$$= \frac{p}{q} \cdot \left(\frac{q}{(q-1)^2} \right),$$

where the computer algebra system Mathematica is used to evaluate the sum. Continuing, we have

$$E = \frac{p}{p^2}, \text{ since } q - 1 = -p$$

$$= \frac{1}{p}$$

$$= \frac{36}{10},$$

or 3.6 rolls.   ■

With no edge on either side, where does the casino make its money? Basic bets at Barbooth pay off at less than even money. As in baccarat, a 5% commission is charged on all bets and collected on losing bets. This is accomplished by paying off bets at 20 to 21 odds: players risk $21 but are paid only $20 if they win. At the same time, casino barbooth offers side bets, similar to the wagers found on a craps table, that also charge the 5% commission. The craps bet "Any 7" is often available, as are separate bets on the eight decisive rolls.

Chuck-A-Luck and Sic Bo

We noted in Section 3.2 that chuck-a-luck's high HA of 7.9% might be responsible for its disappearance from casinos. Assuming the truth of this hypothesis, game designers might reasonably set out to modify the game and reduce the house advantage in order to attract players. There is a delicate balancing act at work here, in that changing the game so much that the gambler has an advantage will result in a game that casinos will not offer.

One possibility would be to introduce a "bonus die" that pays out twice what the other dice do. Suppose that we replace the three identical dice in the standard cage with two white dice and one purple die, with the purple die paying off at 2 to 1 if it shows the player's number.

In this game, the probability of losing $1 remains

$$P(-1) = \left(\frac{5}{6}\right)^3 = \frac{125}{216}.$$

The remaining probabilities change, depending on whether or not the chosen number appears on the purple bonus die. For a $1 win, the number cannot appear on the bonus die, so

$$P(1) = \left[2 \cdot \frac{1}{6} \cdot \frac{5}{6}\right] \cdot \frac{5}{6} = \frac{50}{256},$$

where the term in brackets computes the probability of rolling the chosen number on exactly one of the two white dice, and the remaining 5/6 factor accounts for not rolling the number on the bonus die.

For $2, there are two cases to consider: either the number appears on both white dice or only on the purple die. These cases may be combined using the First Addition Rule, so

$$P(2) = \left[\frac{1}{6} \cdot \frac{1}{6} \cdot \frac{5}{6}\right] + \left[\frac{5}{6} \cdot \frac{5}{6} \cdot \frac{1}{6}\right] = \frac{5 + 25}{216} = \frac{30}{216}.$$

A $3 win occurs when one white die and the purple die show your number, and the third die does not.

$$P(3) = 2 \cdot \left(\frac{1}{6}\right)^2 \cdot \frac{5}{6} = \frac{10}{216}.$$

Finally, you win $4 when all three dice show your number. This is identical to the case of winning $3 at standard chuck-a-luck and so has probability $\frac{1}{216}$.

The resulting probability distribution for the random variable X that counts the winnings on a $1 bet on a single number is shown in Table 4.7.

TABLE 4.7: PDF for chuck-a-luck with a single bonus die

x	-1	1	2	3	4
$P(X = x)$	$125/216$	$50/216$	$30/216$	$10/216$	$1/216$

We note that the probability of losing $1 has not changed. The probability of winning $1 has fallen by $\frac{25}{216}$, with that fraction being distributed to higher payoffs, so the expected value of a bet will increase. The expected return can be calculated to be $.088—so we have overcorrected and created a game with a positive player expectation of 8.8%.

There are, of course, other ways to modify chuck-a-luck. If we just swap out the standard dice for eight-sided dice, we can easily see that the HA for three identical dice will increase, since with more sides, the probability of winning will decrease while the payoffs will still be $1 to $3. What if we incorporated the bonus die idea from above, but with d8s?

With one bonus die, paying $2, and two regular dice, paying $1 each, it turns out that there are too many losing combinations. The probability distribution is shown in Table 4.8 and the expected value comes out to $-$.16992,

TABLE 4.8: PDF for chuck-a-luck with 3d8 and a bonus die

x	-1	1	2	3	4
$P(X = x)$	$343/512$	$98/512$	$56/512$	$14/512$	$1/512$

about a 17% HA—worse than regular chuck-a-luck.

So we try again. Table 4.9 shows the effect of switching to one even money die and two that pay at 2 to 1.

TABLE 4.9: PDF for chuck-a-luck with 3d8 and 2 bonus dice

x	-1	1	2	3	4	5
$P(X = x)$	$343/512$	$49/512$	$98/512$	$14/512$	$7/216$	$1/512$

Things look better; the expectation for this game is $-\frac{23}{512} \approx -$.045$. With a 4.5% HA, we have found a modification of chuck-a-luck that—from a purely mathematical standpoint—might be competitive against roulette and most of the center bets in craps.

The success or failure of a new game, of course, is a function of many other variables, most of which are not purely mathematical.

Sic bo is a three-die game that expands on chuck-a-luck. The chuck-a-luck option of betting on a single number from 1 to 6 remains, and it is joined by a host of additional betting options. For example, one may bet on various propositions involving the sum of the three dice, the occurrence of a specific pair of numbers, or on whether or not the dice will all show the same number. Figure 4.4 shows a betting layout for sic bo. The bottom two lines of the layout are home to the single-number chuck-a-luck bets. At some casinos, including

FIGURE 4.4: Sic bo betting layout [124].

the Mohegan Sun in Montville, Connecticut, the layout is electrified, and the sic bo dealer need only press buttons corresponding to the numbers on the three dice to light up all of the winning bets.

Example 4.2.4. A new option in sic bo is a bet on a specified pair of different numbers, which pays off at 6 to 1 if both numbers appear among the three dice. Among the 216 ways for three dice to land, how many show a given pair of numbers, and what is the expectation of this bet?

The two numbers selected are immaterial, so we shall focus on the 1–2 bet. Designating the dice by Red, Green, and Blue, we have the following winning combinations:

Red	Green	Blue
1	2	Any
2	1	Any
1	Any	2
2	Any	1
Any	1	2
Any	2	1

Since each of the "Any" slots may be filled by six numbers, it appears that there are 36 winning combinations, which would give the player a 16.67% advantage. It follows, therefore, that we have missed something with this model. Certain winning combinations where "Any" is replaced by a 1 or 2 have been counted twice in this accounting; these are 1-2-1, 1-2-2, 2-1-1, 2-1-2, 1-1-2, and 2-2-1. Removing the duplicate occurrences of these six combinations from the list of 36 leaves 30 winning rolls and 186 losing rolls for the 1-2 bet; the resulting expectation on a $1 bet is

$$E = (6) \cdot \frac{30}{216} + (-1) \cdot \frac{186}{216} = -\frac{6}{216} \approx -\$.0278,$$

which gives a house advantage of 2.78%, considerably better than the edge of 7.87% on single-number chuck-a-luck bets.   ∎

Example 4.2.5. Galileo's work in determining the correct probabilities for rolling sums of 9–12 on three dice is essential to sic bo bets on those sums. In Figure 4.4, we see that bets on a sum of 9, 10, 11, or 12 all pay off at 6 to 1. Which of these has the lowest HA?

Let x be the number of ways to roll the specified number. The expectation of a $1 wager is then

$$E(x) = (6) \cdot \frac{x}{216} + (-1) \cdot \frac{216 - x}{216} = \frac{7x}{216} - 1.$$

For wagers on 9 or 12, $x = 25$, and thus the expected return is –$.190, for a 19% house advantage. You do slightly better by wagering on 10 or 11, where $x = 27$ and the HA is 12.5%.   ∎

Due to an unintentional modification, the most exciting day in the history of sic bo was October 26, 1994 [150]. On that day, the Grand Casino in Biloxi, Mississippi was found to be unwittingly offering an 80 to 1 payoff for winning bets that the sum of the dice would be 4 or 17. Since there are three ways to roll a total of 4 on three dice (1-1-2, 1-2-1, and 2-1-1), the expected value of a $1 bet on 4 was

$$E = (80) \cdot \frac{3}{216} + (-1) \cdot \frac{213}{216} = \frac{27}{216} = \$.125$$

—*so players had a 12.5% advantage.* The same player advantage held for a bet on 17.

News of this unusual opportunity spread through the gambling community, largely via fax—this was in the infancy of the World Wide Web—and on the 26th, gamblers from across America converged on Biloxi, including one Minnesota man who drove for 17 hours to reach the casino. They played the game for hours, crowding the table so deeply that it was difficult to find a place. After the game shut down for the night, casino officials estimated that the casino's losses were in the neighborhood of $180,000. Casino management closed the game down the next day, and when it reopened, the payout on 4 and

17 had been reduced to the standard 60 to 1, resulting in a new expectation of

$$E = (60) \cdot \frac{3}{216} + (-1) \cdot \frac{213}{216} = -\frac{33}{216} = -\$.1528,$$

or an HA of about 15.3%.

4.3 Card Games

Card Craps

In California, state law holds that dice cannot be used as the sole device determining the outcome of a game of chance. A similar law held in Oklahoma until the state's compact with its tribal casinos was amended to allow ball and dice games in 2018 [15].

Prior to the change in the compact, a game developed by the Quapaw tribe of Oklahoma used two rows of six cards, each running from ace through 6 and randomly dealt facedown to the table. Players called out two numbers, which determined which cards from each row would be turned over and added to simulate a roll of dice. When the two rows of cards are shuffled and replaced after each simulated roll, this game is mathematically identical to standard craps [45].

California casinos may legally offer a game called *card craps*, a game similar to craps in which a special deck of cards is used to mimic a pair of dice. One version of card craps differs in both the number of cards in use and the probabilities of the various outcomes: either 264 or 324 cards are used, evenly distributed among the numbers 1 to 6. Two cards are dealt to simulate a roll of two dice. The deck is not dealt down very far without replacement; cards are immediately reinserted into a continuous shuffling machine that replaces them into the deck, but there may be a small time lag during which the numbers from 1 to 6 are not exactly equally likely. Specifically, cards that have just been dealt—perhaps as many as 12—are unavailable because of how the machine operates: these cards are held in a buffer and cannot be immediately redrawn so that the machine can dispense cards as needed without an undesirable time lag.

One difference that is immediately apparent and that does not depend on any buffered reserve of cards is that the probability of "rolling" doubles is slightly less than $\frac{1}{6}$, because once the first card is drawn, the six numbers are no longer equally likely for the draw of the second card. Specifically, if a 264-card deck (44 of each rank) is used, the probability p of doubles is the probability that the second card matches the first in rank:

$$p = \frac{43}{263} \approx .1635 < \frac{1}{6} \approx .1667,$$

and with 324 cards (54 of each rank),

$$p = \frac{53}{323} \approx .1641 < \frac{1}{6}.$$

On a more practical level, the use of a continuous shuffling machine with a delay may make the Don't Pass bet more lucrative, with a positive expected value that favors the player who counts cards and adjusts odds bets with the count, taking them down when the count is bad and restoring them when the count is favorable [53]. Remember that odds bets on the Don't Pass line may be removed at any time before being resolved. Suppose, for example, that the point is 4. If a lot of 1s and 2s are dealt out of the deck and held in the buffer, they're unavailable to be dealt immediately, and so the cards will tend to generate higher totals: 7 will be even more favored relative to 4 than it is normally.

EZ Baccarat

The baccarat requirement that winning bettors on the Banker hand pay a 5% commission is a source of some confusion to players and dealers alike. *EZ Baccarat* is an effort to eliminate this confusion by a rule change that changes the edge on a Banker bet from the bettor to the casino: If the Banker hand is a three-card total of 7, then a Banker bet pushes rather than wins against a lower Player hand. A Player wager still loses against a three-card 7 if the Player hand is 0–6.

The odds of a three-card 7 are approximately 43.4 to 1 [25]. This change in the rules, turning a winning hand into a push, is enough to give the house a 1.02% advantage when winning Banker bets are paid at 1 to 1 [64]. As a result, the need to pay commissions is eliminated, as is (more importantly) the time spent collecting them. With an increase in game pace, the casino can deal more hands per hour and so increase its long-term gain.

Often appearing together with EZ Baccarat is the *Dragon 7* side bet, which is a simple bet that the Banker hand will win with a three-card 7 and which pays 40 to 1. The expectation of this wager is

$$E = (40) \cdot \frac{1}{44.4} + (-1) \cdot \left(1 - \frac{1}{44.4}\right) = -\frac{3.4}{44.4} \approx .0766,$$

giving the house a 7.66% advantage. In light of the fact that the HA on the main game is less than 1.25% on either Player or Banker, this is a bet to be avoided, though it is better than the Tie bet.

Casino War

Casino War is a gambler's version of the children's card game War. Player and dealer are each dealt one card from a six-deck shoe (the number of decks may vary), and the high card wins. Aces are always high. At this point, the game is even—both sides have an equal chance of drawing the higher card.

The casino's edge comes from how ties are handled. If the two cards match, players may either *surrender*, forfeiting half their wager and ending the game right away, or, as in the children's game, *go to war*. If war is declared, the player must double his or her bet, and the dealer then burns three cards (removes them from play without exposing them) and deals a second card to each side. If the player's card is higher, the player wins the second bet, which is paid off at 1 to 1 odds, and the first bet is a push; if the dealer's card is higher, the player loses both bets.

A tie on the second card is handled differently at different casinos. A common resolution is to pay the player a bonus equal to the amount of the original bet while the actual bets are declared pushes, a net win of 1 unit. On a $1 bet, then, we have the following player outcomes:

Result	Win/Loss
Win without war	+$1
Lose without war	−$1
Surrender	−$.50
Win after war	+$1
Lose after war	−$2
Tie after war	+$1

The source of the casino's advantage is clear from this chart: In the event of a casino win after a tie, 2 units are won, while the player can never win more than the amount of his or her original wager even if 2 units are at risk in a war.

It may be stretching the definition of the word to speak of a "strategy" for a game as simple as Casino War, but the question of whether to surrender or go to war after a tie is a place where player choice is involved, and thus a place where an optimal strategy may be determined.

If you surrender a $1 bet, your expectation is −$.50. If you go forward with the war, then the probability of winning is again dependent on the number of decks in use. You will either win $1, whether through winning the war or tying, or lose $2 when the war is lost. Let p be the probability of a tie; it follows that

$$P(\text{Win } \$1) = \frac{1}{2} + p$$

and

$$P(\text{Lose } \$2) = \frac{1}{2} - p.$$

We assume no knowledge about the cards remaining to be dealt, and again assume that we're starting at the top of a fresh n-deck shoe. Three cards are known at the start of the war: the two matching cards that triggered the war and the player's second card. The burn cards are not exposed when dealt, so we ignore them—the result will be another long-term average value that is suitable for quick calculations like this one. Two cases emerge:

- If your second card matches the first two in rank, then the conditional probability of a fourth card of that rank falling to the dealer is

$$p_1 = \frac{4n - 3}{52n - 3}.$$

  This case has probability

$$q_1 = \frac{4n - 2}{52n - 2}.$$

- If your second card is of a different rank than the first two, then the conditional probability of a second match is

$$p_2 = \frac{4n - 1}{52n - 3},$$

  and this case has probability

$$q_2 = \frac{48n}{52n - 2}.$$

The probability p of a tie and the expectation E are both functions of n. We find that

$$p(n) = q_1 p_1 + q_2 p_2 = \frac{208n^2 - 68n + 6}{(52n - 2)(52n - 3)},$$

where n is the number of decks in the shoe. Hence, your expectation as a function of n is

$$E(n) = (1) \cdot \left(\frac{1}{2} + p(n)\right) + (-2) \cdot \left(\frac{1}{2} - p(n)\right) = \frac{-36 \cdot (26n - 3)}{(26n - 1)(52n - 3)} - \frac{7}{26}.$$

For the commonly used values of n, Table 4.10 contains the expectation if you go to war. Once again, a limiting value for the case of infinitely many decks can be computed, and here that limit is $-\$\dfrac{7}{26} \approx -\$.269$.

Since the expected value for any number of decks is greater than -50¢, it follows that you should *never* surrender, and thus that the casino derives an additional advantage whenever a player surrenders.

Casino War also offers a Tie bet, which pays 10 to 1 if the first two cards match. Assuming a single hand dealt from a fresh shoe, the HA for this bet depends on the number of decks in play. We shall compute this house edge for a variable number of decks, denoted by n; substitution of commonly used values of n will then give the HA appropriate to any particular game.

Once the first card is drawn, we seek the number of cards remaining in the shoe that match its rank. In an n-deck game, there are $4n - 1$ such cards.

TABLE 4.10: Expected return when going to war, Casino War

n	$E(n)$
1	$-\$.321$
2	$-\$.296$
4	$-\$.282$
6	$-\$.278$
8	$-\$.276$

Since the entire shoe holds $52n$ cards and one (the player's card) has been removed, the probability of a match is

$$p = \frac{4n - 1}{52n - 1}.$$

For a single-deck game, $p = \dfrac{1}{17}$; as n increases, this probability approaches $\dfrac{4}{52}$, or $\dfrac{1}{13}$. The casino's advantage comes from paying 10 to 1 on what is no better than a 12 to 1 shot. The exact HA may be found from the equation

$$E(n) = (10) \cdot \frac{4n - 1}{52n - 1} + (-1) \cdot \left(1 - \frac{4n - 1}{52n - 1}\right) = -\frac{8n + 10}{52n - 1}.$$

The expected values are tabulated for the various values of n typically used in casinos in Table 4.11.

TABLE 4.11: House edge on the Tie bet for Casino War

# of decks	HA
1	35.29%
2	25.24%
4	20.29%
6	18.65%
8	17.83%

It is clear that you should not make this bet, but it is of interest to notice that the casino's edge actually decreases with the number of decks in the shoe, which is the opposite of the effect in a game like blackjack. Increasing the number of decks past eight (the maximum typically in use in casino table games) cannot, however, eliminate the casino advantage completely. In the limit as $n \to \infty$, the expectation approaches

$$E = (10) \cdot \frac{1}{13} + (-1) \cdot \frac{12}{13} = -\frac{2}{13},$$

which corresponds to approximately a 15.38% house edge.

Wild Cards

A common feature in casual poker games, and an occasional component of some video poker machines, is the use of *wild cards*, one or more cards in a deck that may be redefined as any card the holder desires. Common choices for wild cards are a joker added to the deck, deuces, and one-eyed jacks (the $J\heartsuit$ and $J\spadesuit$).

The incorporation of wild cards into poker has a number of effects on the probabilities of the game. One possibly unintentional effect is that, in games with a wild card, three of a kind becomes more common than two pairs. The reason for this is contained in the relative rankings of the two hands: if a player is dealt, say, $A\spadesuit$ $A\diamondsuit$ $5\diamondsuit$ $2\diamondsuit$ and a wild joker, the joker could be regarded as a third ace or as a second 5, and counting the hand as three aces is better than calling it two pairs. In short, there are no two-pair hands that include a wild card.

Example 4.3.1. Suppose we add a single joker to the deck of 52 to function as a wild card. A hand that does not contain a joker will be called a *natural* hand to distinguish it from a hand using a wild joker. The number of natural two-pair hands (see Example 2.1.11) is

$$\binom{13}{2} \cdot \binom{4}{2}^2 \cdot 44 = 123,552.$$

This number will not increase with a wild joker, for the reason given above. Let us count next the number of three-of-a-kinds without a wild card. The rank of the triple may be chosen in 13 ways, and the three cards comprising the triple may be chosen in $\binom{4}{3} = 4$ ways, for a total of 52 triples. Another way to think of this is that there's one three-of-a-kind corresponding to every card in the deck: the other three cards of that rank.

The two remaining cards must not match either the triple or each other; we may choose them in $\dfrac{48 \cdot 44}{2} = 1056$ ways. We divide by 2 because, again, the order in which the two odd cards are chosen does not matter: if we have three kings, a jack, and a 7, whether we pick the jack first and then the 7, or the other way around, the resulting hand is the same.

The total number of natural three of a kinds is thus

$$\frac{52 \cdot 48 \cdot 44}{2} = 54,912.$$

To this we must add the number of three of a kinds which include a joker: these hands are of the form $Wxxyz$, where the W denotes the wild joker. The joker can be dealt in only one way, so we are simply counting the number of four-card one-pair hands here. There are

$$\binom{13}{1} \cdot \binom{4}{2} \cdot \binom{12}{2} \cdot \binom{4}{1}^2 = 82,368$$

three-of-a-kinds including a joker. Adding these to the natural three-of-a-kinds gives 137,280 three-of-a-kinds in a 53-card deck with one wild joker—more than the number of two-pair hands. ∎

In theory, then, the ranking of hands in 53-card poker should be rearranged to place two pairs above three-of-a-kind. In practice, of course, this would simply result in players opting to define a hand including a pair and a joker as two pairs rather than three of a kind—and so negate the mathematical correctness of the new hand ordering. John Scarne suggested the following two possible changes to the rules of poker, for players interested in strict adherence to mathematical laws [104]:

1. In a hand consisting of a pair plus a joker, the joker cannot be called wild and must be valued as a third unmatched card.

2. Alternately, in a hand with pair of 8s through aces plus a joker, the joker may be used to raise the hand to three-of-a-kind. A pair of 2s through 7s plus a joker must be called two pairs.

Under either one of these alternate hand-ranking schemes, three of a kind is less common than two pairs. In suggesting these rule changes, Scarne was well aware that his hybrid schemes might be mathematically correct, but would be very unlikely to catch on for casual play. The first scheme simply eliminates all two-pair and three-of-a-kind hands containing a wild card. Six ranks of pairs cannot be raised to three of a kind by a joker in the second scheme, while seven ranks can. This discrepancy does not overcome the excess of natural two-pair hands over natural three-of-a-kinds, and so the order of the two hands is preserved.

Example 4.3.2. As an example of the challenges that arise when playing poker with wild cards, we will show that in a 54-card deck with 2 wild jokers, full houses and four-of-a-kind hands are equally likely.

Natural hands of both types can be counted using the formula on page 44: there are 624 four-of-a-kind hands and 3744 full houses without jokers.

To form a four-of-a-kind hand with 1 joker, the hand must be $Wxxxy$, where x and y are different card ranks—which could also be called a full house, but isn't because four-of-a-kind beats a full house with a standard deck. There are 2 choices for the joker, 13 for the rank of the triple, $\binom{4}{3} = 4$ ways to pick the cards of the triple, and 48 ways to choose the odd card of rank y. Multiplying gives 4992 one-joker four-of-a-kinds.

A full house with a single joker looks like $Wxxyy$, and the joker is used as the higher rank of x and y. These hands number $2 \cdot \binom{13}{2} \cdot \binom{4}{2}^2 = 5616$.

Hands with 2 jokers become four-of-a-kinds if the hand is $WWxxy$, with $x \neq y$. There is only 1 way to select the jokers, and then $13 \cdot \binom{4}{2} \cdot 48 = 3744$ ways to fill out the rest of the hand.

There are no full houses with 2 jokers, since any such hand would contain a natural pair of xs and would thus be played as four of a kind. These hands look like $WWxxy$, and there are $13 \cdot \binom{4}{2} \cdot 48 = 3744$ of them.

Adding everything up gives 9360 of each type of hand. ∎

The implications of this result for gameplay are left to individual poker players. It is important for all players to agree on how jokers may be used when playing poker with wild cards.

A later examination of the mathematics by John Emert and Dale Umbach considered poker hands dealt from a deck with one, two, or four wild cards [21]. This analysis showed that if wild cards are allowed in poker, it is not possible to rank the hands so that more valuable hands occur less frequently without restrictions on how wild cards may be valued. No matter how the various hands are ranked, there will always be a way to value hands containing wild cards so that a more common hand is ranked higher than one that is less common.

In 2011, Kristen Lampe proposed the *Wild Card Rule* as a way to preserve hand rankings while incorporating one or more jokers into poker [66].

> **Wild Card Rule**: A poker hand containing one or more wild cards is assigned its *second highest* possible rank, using the standard poker hand rankings as an ordering.

Here are some applications of the Wild Card Rule:

1. In a 53-card deck with 1 wild joker, a hand only counts as three-of-a-kind under the Wild Card Rule if it contains no wild cards. Three of a kind is a possible hand valuation with a joker only if the hand also contains a natural pair. The hand $W\ 7\heartsuit\ 7\clubsuit\ A\diamondsuit\ 9\diamondsuit$ can be valued as three-of-a-kind, 2 pairs, or 1 pair. The Wild Card Rule would call this 2 pairs. This is how any hand of the form $Wxxyz$ will play.

   A hand of the form $Wxxxy$ can be scored as either three-of-a-kind, a full house, or four-of-a-kind. Since the full house ranks second, the hand is called a full house.

2. If a game uses 2 or more wild cards, the hand $W\ W\ K\clubsuit\ J\clubsuit\ J\spadesuit$, with 2 wild cards, has multiple evaluation possibilities. This hand can be four-of-a-kind, a full house, three-of-a-kind, 2 pairs, or 1 pair. The second-highest option is a full house: kings over jacks.

3. $W\ W\ J\diamondsuit\ 6\diamondsuit\ 3\clubsuit$ can be counted as three-of-a-kind, 2 pairs, 1 pair, or high card. The Wild Card Rule directs that this hand play as 2 pairs.

   If the $3\clubsuit$ is replaced by the $3\diamondsuit$, the new hand can also be counted as a flush, and so is elevated to three-of-a-kind.

4. Any hand of the form $Wxyzw$, with no flush or straight possibilities, cannot be promoted to one pair and must be valued as a high-card hand.

5. A desirable corollary to the Wild Card Rule is that there is no unbeatable hand, even with a wild card in play. There are no five-of-a-kind hands using the Wild Card Rule. Since five-of-a-kind ranks highest, no hand containing a wild card can have five-of-a-kind as its second-highest valuation.

One unexpected challenge to the Wild Card Rule arises with a hand like $W\ K\Diamond\ 9\Diamond\ 5\Diamond\ 3\Diamond$. The possible assignments of W give a flush, a pair, or a high-card hand. The Wild Card Rule ranks this hand as a pair of kings. If the game is draw poker, a player might be tempted to discard the joker in the hope of improving the hand to a natural flush [66].

By discarding the wild card and drawing, the following hands, shown with their probabilities, are possible:

$$P(\text{Flush}) = \frac{9}{48} = \frac{3}{16}.$$
$$P(\text{Pair}) = \frac{12}{48} = \frac{1}{4}.$$
$$P(\text{High card}) = \frac{27}{48} = \frac{9}{16}.$$

The probability of improving this pair of kings by discarding the joker is only $\frac{3}{16}$. There is a $\frac{1}{16}$ chance that the hand remains a pair of kings, and the probability that the hand declines in value is $\frac{3}{4}$.

A better strategy might be to discard three cards, keeping the king and joker, or just to hold the joker and draw four new cards.

A variation on this idea may be found on some video poker machines, where a royal flush with wild cards pays off somewhat less than a natural royal flush, but video poker has the advantage of being automatically scored and paid off, and unlike live poker need not be concerned with human interpretation of its results. One pay table for a video poker machine dealing from a 53-card deck including one wild joker is Table 4.12 [59].

The big payoff—indeed, the only inflated payoff at the five-coin level— remains the natural royal flush. A royal flush with a joker pays well, but not nearly as well as the natural royal, and it collects no bonus when betting max coins.

An additional modification that appears in this table is the fact that two pairs is now the lowest paying hand, although in the play of the game, a pair plus a joker is interpreted as three of a kind. Some joker-wild games pay off on a pair of kings or better; this usually comes along with slightly lower payoffs further up the table.

TABLE 4.12: Video poker pay table with one wild joker

Poker hand	Payoff: 1 coin	Payoff: 5 coins
Natural royal flush	500	4700
Five-of-a-kind	100	500
Joker royal flush	50	250
Straight flush	50	250
Four of a kind	20	100
Full house	8	40
Flush	7	35
Straight	6	30
Three of a kind	2	10
Two pair	1	5

An obvious question arises: If you are dealt a joker royal flush, is it a good idea to discard the joker and go for the natural royal? The probability of hitting it is 1/48, but the payoff is greater by a factor of 18.8 when playing max coins, and there are smaller payoffs possible to fill the gap.

This problem is similar to the one explored in Example 3.2.28; with a similar table of probabilities. Assume that you have been dealt $KQJT\heartsuit$ along with a joker. On a $1 machine with max coins ($5) bet, your possible outcomes if you discard the joker are these:

Result	Net Payoff	Probability
Natural royal flush	4695	1/48
Straight flush	245	1/48
Flush	30	7/48
Straight	25	6/48
Nothing	−5	33/48

The expectation if you hold the joker is $245. If you discard the joker, your expectation is

$$E = (4695) \cdot \frac{1}{48} + (245) \cdot \frac{1}{48} + (30) \cdot \frac{7}{48} + (25) \cdot \frac{6}{48} + (-5) \cdot \frac{33}{48} \approx \$106.98.$$

It's not even close. As one might expect, breaking up a joker royal flush is not a good strategy, and the corresponding strategy table for this game makes that clear: A joker should never be discarded [59].

A second option seen in video poker is *Deuces Wild*, in which all four 2s function as wild cards. In the long run, of course, this leads to higher-ranked hands, and the pay table must accommodate that. On a Deuces Wild machine, payoffs for a pair of jacks or better and for two pairs are typically eliminated, and the lowest hand that pays off is three of a kind. Table 4.13 shows a Deuces Wild pay table with a 100.76% return [117].

TABLE 4.13: Full pay Deuces Wild pay table with 100.76% return [117]

Poker Hand	Payoff: 1 coin
Natural royal flush	800
Four deuces	200
Wild royal flush	25
Five of a kind	15
Straight flush	9
Four of a kind	5
Full house	3
Flush	2
Straight	2
Three of a kind	1

Example 4.3.3. There are four natural royal flushes. In each hand, we can replace up to four cards by 2s and still have a royal flush. What is the probability of a dealt wild royal flush in Deuces Wild poker?

We proceed by counting the number of 2s. If there is one deuce in the hand, there are five cards that it can replace, and four choices for the replacing card. Multiplying by the four natural royal flushes gives a total of $4 \cdot 5 \cdot 4 = 80$ one-deuce royals.

If our royal flush contains two deuces, we can pick the deuces in $\binom{4}{2} = 6$ ways, and the exiting cards in $\binom{5}{2} = 10$ ways. Across all four natural royals, this leads to $4 \cdot 6 \cdot 10 = 240$ royal flushes with two wild cards.

A three-deuce royal has four choices for the deuces and $\binom{5}{3} = 10$ ways to pick the replaced cards. There are therefore $4 \cdot 4 \cdot 10 = 160$ such hands.

Four-deuce royals are easy to count: We simply combine the four deuces with any of the 20 cards that can be part of a royal flush, for a total of 20.

Adding everything up gives a total of 504 royal flushes, 126 times the number without wild cards, and so the probability of a royal flush is also 126 times greater, or

$$\frac{504}{2,598,960} = \frac{3}{15,470} \approx 1.939 \times 10^{-4}.$$

∎

This plethora of new royal flushes is reflected in the pay tables, where a natural royal flush pays anywhere from 10 to 32 times as much as a royal flush with one or more deuces, depending on the exact game being played. Of course, each pay table carries its own perfect strategy, and players using the wrong strategy will increase the often-tiny house advantage.

4.4   Casino Promotions

We have seen the Baldini's promotion (Example 3.2.4) of a free spin on a video poker machine as an incentive for local patrons to cash their paychecks at the casino. Another incentive for players comes in the form of *matchplay* coupons, which allow players to make a bet at a table game and have the amount they wager matched, in principle, by the casino. No additional chips are played; the coupon functions as a chip.

The net effect of a matchplay coupon is to double the payoff, should the bet win. Typically, these coupons are restricted to even-money wagers and have some maximum bet limit. A matchplay coupon offered by the Downtown Grand Casino of Las Vegas in 2022 effectively offers a 2 to 1 payment on any even money bet of $50 or less by doubling the amount of a player's wager. The coupon must be played with a live bet of the amount indicated and is surrendered after the bet is either won or lost; in the case of a tied bet, as in blackjack, the player may re-bet the coupon. It is not unusual for this increased payoff to tip the game to the player's advantage, hence the bet limit and restrictions on the coupon's use.

Example 4.4.1. For a $50 matchplay bet on red at roulette at the Downtown Grand, what is the player's advantage?

The probability of winning has not changed; what is different is the payoff to a winning player, which has doubled. In American roulette, we have

$$E = (100) \cdot \frac{18}{38} + (-50) \cdot \frac{800}{38} = \frac{160}{38} \approx \$21.05,$$

which, when divided by the $50 initial bet, gives the player a 42.1% edge over the casino. ■

If this coupon could be used on a 35 to 1 single number bet in roulette, a payoff of $1750 would be doubled to $3500. Accordingly, the expectation would be

$$E = (3500) \cdot \frac{1}{38} + (-50) \cdot \frac{37}{38} = \frac{1650}{38} \approx \$43.42,$$

for a player advantage of 86.8%.

Related to matchplay coupons are *nonnegotiable* or *no cash value* (NCV) chips. These special chips have a face value, but may not be redeemed at the casino for cash. Instead, they must be played at the tables at least once. They need not be accompanied by matched cash bets, and if a bet made with these chips wins, the player is paid with ordinary negotiable casino chips. NCV chips come in two types. The first type, and the most common, may be wagered only once, and like a coupon, is taken after a resolved bet—win or lose. The second type, while occasionally offered to any casino patron, is usually available only to a casino's best customers, and can be retained after a win to bet again, until they lose.

Both forms of nonnegotiable chips are, like matchplay coupons, typically restricted to even-money wagers. One-use chips have an expected value of approximately half their face value, because they are only in play for a single wager. If an NCV chip has a face value of $\$A$ and is wagered on an even-money bet whose probability of winning is $p < .5$, it follows that

$$E = (A) \cdot p + (0) \cdot (1 - p) = A \cdot p \approx \frac{A}{2},$$

where the expectation is roughly as close to $\frac{A}{2}$ as p is to ½. For even-money bets, p is usually very close to, but always less than, ½.

Bob Stupak took this consideration and turned it around on his customers at Vegas World. Stupak was known in Las Vegas for his "VIP Vacation" packages, which he liberally offered and aggressively marketed in order to attract visitors to Vegas World, which was located on Las Vegas Boulevard about half a mile north of Sahara Avenue, the traditional northern boundary of the Las Vegas Strip, in a neighborhood of questionable reputation known as "Naked City." The packages offered a low-cost Las Vegas vacation that included two nights at the resort, complimentary drinks, and free gambling money in the form of cash and special gambling chips, as much as $1000 in chips per visit, all for less than the face value of the chips alone. These not-quite-NCV chips, shown in Figure 4.5, were inscribed with a table value and a cash redemption value of ¼ that value, so a $25 chip had a cash value of $6.25 and a $5 chip had a cash value of $1.25. Vegas World guests who were

FIGURE 4.5: Vegas World promotional chips. Each chip had cash redemption value equal to 25% of its face value.

unfamiliar with gambling mathematics and were happy to cash in their chips for their redemption value were handing the casino nearly 25% of the chips' face value, for, as was shown above, the chips had an expected return of close to half their face value.

An alternate Vegas World promotion issued patrons two $100 bills when they checked in, with instructions to take them to the casino cashier to exchange them for $600 in nonnegotiable chips [125]. Customers, of course, were free to disregard the instructions and keep the money; anyone who did so was essentially forfeiting nearly $100 in value. Many did.

Example 4.4.2. What would be a good way to use those Vegas World chips? As we will see in Section 6.5, maximizing your return in a casino is best done by making a few large bets rather than many small ones. Several people independently made the following discovery: if you have $1000 in non-negotiable chips, take them to a roulette table and bet $500 on red, bet the other $500 on black, and put $20 in regular casino chips on each of the zeroes. You have every possibility covered, and so cannot lose. The outcomes of this wagering combination are these:

- **A red or black number comes up.** You lose your bets on the zeroes (−$40), but win $500 in casino chips on your color bet. Your NCV chips are taken, but you can walk away with a $460 profit.

- **0 or 00 is spun.** You lose all of your NCV chips, but win $700 on your winning bet, from which we subtract the $20 you wagered on the losing number, for a net profit of $680.

Your expected return with this combination bet at an American roulette table is

$$E = (460) \cdot \frac{36}{38} + (680) \cdot \frac{2}{38} \approx \$471.58.$$

This is a guaranteed return of about 99.56% of the $473.68 $\left(1000 \times \frac{18}{38}\right)$ theoretical expected value of $1000 in NCV chips when wagered on an even-money roulette bet. ∎

It's possible to do better. Take your $1000 to a craps table and put $500 on pass, $500 on don't pass, and $20 cash on 12, a one-roll bet paying 30 to 1 if a 12 is rolled. As with the roulette example, every possible roll of the dice leads to a win for you.

- If a 12 is rolled on the come-out roll, your pass line bet chips are taken, but your don't pass chips are still yours to bet again, and your bet on 12 wins $600.

- In any other case, exactly one of your pass and don't pass bets wins, whether by a 2, 3, 7, or 11 on the come-out roll or by the shooter setting and making a point. You win $500 but lose the $20 wager on the 12, for a profit of $480 to take away with you.

Your expectation on the first bet with this wagering strategy is

$$E = (480) \cdot \frac{35}{36} + (600) \cdot \frac{1}{36} \approx \$483.33.$$

The theoretical expectation of $1000 bet this way is $485.50; your return here is again 99.56% of the long-term average return. If the come-out roll was a 12, you can continue by mimicking this wager with the $500 in NCV chips you originally bet on don't pass, which you retain since that bet was a push.

Casino personnel, while not happy to see the NCV chips being used in this way, evidently made no move to stop players from making these bet combinations. Of course, most people were not aware of this potential and made many bets against the intractable house edge.

Example 4.4.3. A nonnegotiable chip that can be re-bet until it loses has an expected value of approximately its face value. This can be seen in the following calculation: Assume once again that the chip has a face value of A and is being wagered on a bet where the probability of winning is $p < .5$, repeatedly until it loses. Since the chip cannot be cashed in, this is standard procedure with this type of NCV chip. We then have

$$E = (0) \cdot (1-p) + (A) \cdot p \cdot (1-p) + (2A) \cdot p^2 \cdot (1-p) + \cdots$$

$$= \sum_{k=0}^{\infty} kA \cdot p^k \cdot (1-p)$$

$$= A \cdot (1-p) \cdot \sum_{k=0}^{\infty} k \cdot p^k$$

$$= A \cdot (1-p) \cdot \left(\frac{p}{(p-1)^2} \right), \text{ as seen in Example 4.2.3.}$$

$$= \frac{A \cdot p}{1-p}.$$

It should be noted that this calculation includes some very unlikely events with very high payoffs, such as winning 1000 straight bets. If $p = .5$, then the expectation above is A; if $p < .5$, then the expectation is less than A. For most even-money casino bets, p is less than, but close to, .5, and so an upper bound for the value of a replayable nonnegotiable chip is its face value. ∎

Table 4.14 collects some common even-money wagers and shows the expectation of a renewable NCV chip when used on these propositions.

Other casino promotions come in the form of games that offer a particularly advantageous wager, but without a matchplay component. As with the sic bo opportunity in Mississippi described on page 184, these may arise if games are placed on the casino floor in error and without correct mathematics backing them up.

Example 4.4.4. A standard $3 bet on a six-spot keno ticket at Jerry's Nugget in North Las Vegas, Nevada has the payoff table shown in Table 4.15.

This game carries a fairly standard 25.40% house edge. In 1990, the Continental Casino in Las Vegas (now the Silver Sevens) offered a six-spot keno game with the pay structure for a $3 bet shown in Table 4.16 [149]:

TABLE 4.14: Expectation of an $A renewable NCV chip

Wager	$P(\text{Win})$	Expectation
American roulette, even money bet	.4737	$.9000 \cdot A$
Blackjack, basic rules	.4750	$.9048 \cdot A$
Craps, don't pass line	.4790	$.9194 \cdot A$
European roulette, even money bet	.4865	$.9474 \cdot A$
Craps, pass line	.4920	$.9685 \cdot A$
Baccarat, player	.4932	$.9732 \cdot A$

TABLE 4.15: $3 Pick 6 keno pay table from Jerry's Nugget

Outcome	Payoff (for 1)	Net win
Match 4	$9	$6
Match 5	$315	$312
Match 6	$7800	$7797

TABLE 4.16: $3 Pick 6 keno pay table from the Continental Casino

Outcome	Payoff (for 1)	Net win
Match 4	$15	$12
Match 5	$800	$797
Match 6	$5500	$5497

Improving the payoffs for the Match 4 and Match 5 wins had a potentially unintended effect on the house advantage which was not balanced by diminishing the payoff on the very unlikely event of catching all 6 numbers. Writing $P(x)$ for the probability of catching x numbers gives the expected return on a $3 bet at this game:

$$E = (15) \cdot P(4) + (800) \cdot P(5) + (5500) \cdot P(6) - 3 = \frac{194,217}{316,316} \approx \$.614,$$

a positive expectation which gives a player edge of 20.47%. ∎

A challenge in fully exploiting this opportunity—should you ever find it again—is that the probability of even a "Match 4" win, $P(4)$, is only about 2.85%, and so there is some risk that your bankroll might not be large enough to continue betting until you win.

Example 4.4.5. In December 1989, the Sahara Casino in Las Vegas ran a baccarat promotion that eliminated the commission entirely [149]. The expectation of a $1 Banker bet was then

$$E = (1) \cdot (.4584) + (0) \cdot (.0955) + (-1) \cdot (.4461) = \$.0123,$$

giving gamblers a 1.23% edge on Banker bets. ∎

Although the promotion ran only from Monday through Thursday, and then only from 7:00 P.M. to 2:00 A.M., the attraction was so great that seats were seldom available at those tables.

4.5 Exercises

Answers to starred exercises begin on page 288.

Wheel Games

4.1.* Find the house advantage of a $1 Colors bet when made at a Sands Roulette table.

4.2. In 2022, a new roulette bet called *2 G's* was introduced [121]. G stands for "Green" here; the 2-spin bet pays 350–1 if the next 2 spins after the bet is made result in a green number.

a. Find the HA of this bet if made on an American roulette wheel.

b. At 350–1, this bet would have a high player advantage if it were offered at Sands Roulette. Suppose that the Sands version of 2 G's pays off at x to 1, where x, for dealer convenience, is restricted to a multiple of 5. What value for x gives a house edge that is closest to 5%?

4.3.* In January 2023, the Massachusetts Lottery introduced *The Wheel of Luck*, a roulette-based electronic game that was drawn between rounds of the lottery's electronic keno game. The Wheel of Luck contained 36 numbers–no zeroes–and offered both single-number and 18-number bets.

a. A single-number bet paid off at 25 for 1. Find the house advantage.

b. 18-number bets were available on even, odd, red, and black. Each had probability ½ of winning and paid off at 3 for 2: a $2 winning bet returned $3 in total. What is the HA of an 18-number bet?

4.4. In looking at Table 1.5, the list of roulette bets, we notice that every proper divisor of 36 has a corresponding roulette wager, except for 9. A hypothetical 9-number roulette bet should pay 3–1. One obsolete wager gave the player an opportunity to bet on 9 numbers. By laying a chip on the line separating Low and Even on the layout, a player could make a bet covering both cases. If the number spun was both low and even—and there are 9 of those—the bet paid 1–1. If the number was low or even, but not both, the bet pushed. A number that was neither low nor even meant that the bet lost [94].

Find the HA of this bet on an American wheel.

4.5. Confirm the assertion in Example 4.1 that the house advantage for a $1 bet that the number 23 will come up on both wheels of a European Double Action roulette game is 12.3%.

4.6.* *Multicolore* is a roulette-like game that debuted at Monte Carlo in the 1950s [22, p. 115–6]. A wheel is divided into 25 sectors: six each colored green, red, white, and yellow, and one colored blue. A billiard ball is tossed into the spinning wheel and eventually comes to rest within one of the sectors. Players bet on a color: the blue sector is labeled 24 and the six sectors of each of the other four colors are labeled 4-3-3-3-2-2. A bet on blue pays off at 24 to 1, a bet on any other color pays off at x to 1 in accordance with the number x on the sector.

a. Find the house advantage for a bet on blue.

b. Find the house advantage for a bet on any of the other colors.

c. What is the best betting strategy for Multicolore?

4.7. *Boule* or *La Boule* is another roulette variation that, as the name suggests, is of French origin [98]. The wheel is much simpler, using only the numbers 1 to 9. A rubber ball is thrown into the spinning wheel and eventually settles into a hole at the center. Each number has four holes allocated to it. Numbers 1, 3, 6, and 8 are black; 2, 4, 7, and 9 are red. The 5 is yellow and functions somewhat like the green zero in standard roulette. The following bets are available at La Boule:

Bet	Payoff	Description
Red/Black	1 to 1	Bet on red or black
Low (Manqué)	1 to 1	Bet on 1, 2, 3, 4
High (Passé)	1 to 1	Bet on 6, 7, 8, 9
Odd/Even	1 to 1	Bet on odd or even
Single number	7 to 1	Bet on any one number

The number 5 is neither odd nor even, high nor low. Notice that the payoff structure of La Boule is considerably simpler than standard roulette or Royal Roulette—only two different payoffs are available. Compute the house advantage on each of the two payoffs. How do these HAs compare to those of standard roulette and Royal Roulette?

4.8.* In 2008, New Jersey gaming officials approved a *seven-number* roulette bet, which must be made on the block of numbers 10, 11, 12, 13, 14, 15, and 33. If one of those seven numbers is spun, the bet pays off at 4 to 1 [85]. Find the expected value of a $1 seven-number bet on both single-zero and double-zero wheels.

Dice Games

4.9. The Hollywood Casino in Joliet, Illinois was at one time the only casino offering the "All Day 2" bet, which is a bet that a 2 will be rolled before a 7. There was also an "All Day 12" bet, which is mathematically identical to All Day 2. This bet pays off at 5 to 1. Find the house advantage.

4.10. In 2008, the *7 Point 7* craps side bet debuted at the Orleans Casino in Las Vegas [37]. The bet is made before a come-out roll, and is named for the ways that it wins:

- If the come-out roll is a 7, the bet wins and pays 2–1. 7 Point 7 loses if the come-out roll is 2, 3, 11, or 12.

- If a point is established, the bet remains active and is resolved on the next roll.

- If that next roll is a 7, then the bet wins, paying 3–1. Any other roll loses.

Find the probability of winning and the HA of 7 Point 7.

4.11. Does the player have an advantage if the 7 Point 7 bet is offered at a Crapless Craps table?

4.12. Here's a combination of chuck-a-luck and craps that has been seen in the craps pit at a number of casinos. The player may bet on any single number from 1 to 6 and is paid off according to the number of dice that show that number. If two dice show your number, the payoff is 4 to 1; if one die shows your number, the bet is paid off at 2 to 1. Find the expected value of a $1 bet on a single number. How does this compare with the HA of more traditional craps bets?

4.13. We saw that switching from six-sided to eight-sided dice in chuck-a-luck increases the already high house advantage to an even higher and thus less attractive number. It stands to reason (and can be confirmed mathematically) that switching from six-sided to four-sided dice would give the players too much of an edge, but what if we used nonstandard d4s? Consider a chuck-a-luck cage containing three four-sided dice numbered as follows:

Die	Numbers
Red	1, 2, 3, 4
Green	5, 6, 1, 2
Blue	3, 4, 5, 6

Note that for any number you choose, one of the three dice cannot show it, and so the maximum payoff is $2. Find the house advantage for a $1 bet on a single number.

4.14. The Strat Casino in Las Vegas, successor to Vegas World, continues to offer crapless craps and does so with 10X odds available.

a. Remembering that odds bets are paid off at true odds with no house advantage, find the payoff odds for an odds bet when the point is 2 or 12.

b. Find the payoff odds for an odds bet on 3 or 11.

c. Find the house advantage of a $5 pass line bet backed up with maximum 10X odds.

4.15.* Sic bo also offers the *Big* bet, which pays off at 1 to 1 if the sum of the three dice is between 11 and 17 inclusive, with the provision that the bet loses if triples are rolled. Find the HA of the Big bet.

4.16. The *Any Triple* sic bo bet pays off at 30 to 1 if all three dice show the same number. What is the house advantage of this bet?

Card Games

4.17.* *Five Deck Poker* is a video poker innovation that deals each of the 5 cards in a player's hand from a separate 52-card deck. This greatly increases the number of possible hands, to $52^5 = 380,204,032$. Included among this number are new hands such as five-of-a-kind and suited five-of-a-kinds consisting of 5 copies of the same card. When drawing 1 card to the 4-card straight flush $\diamondsuit6789$, is the chance of completing the straight flush higher in Jacks or Better or in Five Deck Poker?

Omaha is a variant of Texas hold'em where players are dealt 4 hole cards instead of 2 and must use exactly 2 of them to build their final hand. A player dealt all 4 aces as hole cards can only use 2 of them—and will not see any help in the form of additional aces among the 5 community cards.

4.18.* If your hole cards are $K\spadesuit$ $9\heartsuit$ $6\spadesuit$ $6\clubsuit$ and the board shows $K\clubsuit$ $K\diamondsuit$ $T\clubsuit$ $8\clubsuit$ $5\clubsuit$, what is the best 5-card Omaha hand you can make?

4.19.* A Texas hold'em hand can be formed from 7 cards in $\binom{7}{5} = 21$ ways. How many possible Omaha hands can a player make from 4 hole cards and a 5-card board?

4.20.* In a 4-handed game of Omaha, find the probability that 2 or more players tie with a royal flush.

Casino Promotions

4.21.* Another Vegas World promotion was a $100 "Field Roll" chip. Find the expectation of this chip when used on a field bet (see Example 3.2.12). Assume that the chip is taken after the field bet is resolved.

4.22. The Lady Luck Casino in downtown Las Vegas (now the Downtown Grand) once issued a coupon offering a $250 payoff on a winning $5 roulette bet on the number 00. Find the player's advantage when using this coupon.

4.23. Many sports books make their money by requiring that players risk $11 to win $10, or, multiplying by 10, risking $110 to win $100. This locks in a 4.55% HA for the book, provided that action on both teams is approximately equal. Some books run occasional promotions that reduce the required bet from $110 to $105. The casino hopes to make up for the smaller HA that results by booking more bets. What is the HA of a $105 bet under these terms?

4.24. The Golden Nugget in downtown Las Vegas once offered a 20-game "Million Dollar Parlay Card" which offered a $1,000,000 payoff for a $5 bet, with the provision that all ties lost [118].

a. Find the house advantage on this bet for a gambler with 50% accuracy in picking winners.

b. How proficient would a player have to be to make this a break-even wager?

Chapter 5

Blackjack: The Mathematical Exception

Thus far, we have not said much about *blackjack* or *21*, arguably the most popular casino table game and certainly one of the most popular subjects of casino mathematics study. There are good reasons for this—some mathemat-

FIGURE 5.1: Blackjack table [96].

ical, some logistical:

- As mentioned in Section 2.4, successive hands of blackjack are *not* independent, and so the Multiplication Rule in its simplest form is not applicable. In part because of this nonindependence, the probability of winning a hand and the expected value of a bet are somewhat more difficult to calculate.

- The rules of blackjack fluctuate from casino to casino—or even from table to table within a casino, making universal calculations impossible. While the best that we can do is build a collection of probabilities as are

DOI: 10.1201/9781003358183-5

appropriate for different game conditions, this is no barrier to interesting mathematics. It simply means that we must be careful about tailoring our calculations to the game at hand and specifying the applicability of our conclusions.

- Blackjack is a game with a skill component, and so player action and player error must be factored into any calculations.

A novice blackjack player would do well to learn *basic strategy* (Section 5.3), which is a set of rules for decision-making for a player to follow. Based on millions of simulated hands, basic strategy gives a player the best advice on how to play a hand, given the composition of the hand and the dealer's upcard. Calculations about the HA typically assume that the player is using basic strategy correctly; deviations from these instructions will work to increase the casino's long-term edge.

5.1   Rules of Blackjack

A common notion is that the object of blackjack is to get a hand of cards totaling as close to 21 as possible without going over. This is not quite accurate: the object of blackjack is to get a hand that is *closer to 21 than the dealer's hand* without going over.

If the first object listed were the point of the game, stopping on a hand of 12, as is often called for in blackjack basic strategy, would be a bad idea. However, there are times when standing on a relatively weak hand in hopes that the dealer's hand will "bust," or go over 21, is the best strategy for a player.

Basic Rules

Blackjack uses anywhere from one to eight decks of cards shuffled together. Two cards are dealt to each player, and two—one face up, the *upcard* and the other face down, the *hole card*—to the dealer. Players' cards are customarily dealt face down from the dealer's hand in single- or double-deck games, and face up from a shoe in games using four or more decks. Each card counts its face value, with the exceptions that face cards count 10, and an ace may be counted as either 1 or 11, at the player's discretion. (In writing about blackjack, it is common to use the shorthand "ten" to refer to any card—10, jack, queen, or king—counting as 10.)

A hand containing an ace counted as 11 is called a *soft* hand, because it cannot be busted by a one-card draw—if a hand counting an ace as 11 goes over 21 upon drawing a card, the ace may simply be revalued at 1. A hand with a total of 12 to 16 without any aces, or with all aces counted as 1, is called a *hard* or *stiff* hand, because drawing a single card risks busting the hand. A player dealt a two-card total of 21 consisting of an ace and a ten-count card—called a *natural* or *blackjack*—wins immediately unless the dealer also has a natural. Naturals pay either 3 to 2 or 6 to 5, depending on the rules of the casino. (The alternate name "blackjack" for a two-card total of 21 derives from the earliest days of the game, when a 21 consisting of the ace and jack of spades qualified for a bonus payoff.)

If the dealer does not have a natural, then each player in turn has the opportunity to "hit" their hand and take additional cards in an effort to bring their total closer to 21 without going over. If a player's hand exceeds 21, this is called *busting* or *breaking*, and the bet is lost and collected at once.

Once all player hands are settled, the dealer exposes the hole card. If the dealer's hand is 16 or less, he must take additional cards. The dealer must stand on a hand of 17 or higher, although many casinos also require dealers to hit a soft 17 hand. When the dealer's hand is complete—either by busting or reaching a total of 17 or higher—the hand is compared to those of all players who have not yet busted. Player hands that are closer to 21 than the dealer's are paid off at 1 to 1; if the player and the dealer have the same total, the hand is called a *push*, and no money changes hands. If the dealer's completed hand is closer to 21 than the player's, the player loses and his or her wager is collected.

Example 5.1.1. Suppose that two players are facing the dealer in a blackjack game dealt from a single deck. Player 1 is dealt $Q\spadesuit$ $8\clubsuit$ and player 2's hand is $6\heartsuit$ $4\heartsuit$. The dealer's upcard is the $3\heartsuit$.

Player 1 chooses to stand on her total of 18, and player 2 hits his 10, drawing the $2\heartsuit$ and bringing his total to 12. He draws a fourth card, the $T\diamondsuit$, and busts with 22, losing his bet. The dealer turns over the $K\heartsuit$ for a 13, and must draw. His third card is the $2\diamondsuit$, bringing his total to 15. Since this is still less than 17, he draws again and receives the $2\clubsuit$. His total is 17, and he stops. Player 1 wins with her 18. ∎

The advantage for the casino lies in the fact that the players must play out their hands first, and if they bust, they lose even if the dealer subsequently busts with a higher total. The blackjack rule that "ties are a push and the player neither wins nor loses" applies *only* to ties at 21 or less—if a player and dealer tie with hands of 23, the player's chips are already in the dealer's rack before the dealer busts, and they aren't returned.

Additional Rules

Depending on the casino, players may be offered several options during play to make additional bets that offer the chance of winning more money (or, in the case of surrender, losing less money). These are not options available to the dealer.

- If the player's first two cards are the same—as in a pair of 8s or aces—they may be *split* to form the first card of two separate hands. The player must match his bet on the new hand, and the two hands are played out separately. Some casinos allow players to split two 10-count cards, such as a jack and queen, and a standard casino rule allows the player to draw only one additional card to each hand after splitting aces. If a third card matching the first two is drawn to a split hand, many casinos allow that hand to be split again, although some do not allow resplitting of aces. Most casinos have a limit on the number of times a given hand may be split: a maximum of four separate hands is common.

- The player has the option to *double down*—to double his or her initial bet after the first two cards are dealt. This represents a chance for the player to get more money in play upon receiving a good initial hand, but this opportunity comes at a cost: only one additional card may be drawn to a doubled hand. Candidates for double-down hands are hands totaling 9, 10, or 11, as well as certain soft hands. Casinos may place restrictions on which hands may be doubled; some, for example, restrict doubles to 10s and 11s. Some casinos do not allow players to double down after splitting pairs. It is also possible to "double down for less": to increase the bet by less than the full amount originally wagered—this is only recommended if you absolutely cannot afford to double your bet, as doubling for less means taking less than full advantage of a situation where you have the edge over the house.

- If the dealer's face-up card is an ace, players have the opportunity to make an *insurance* bet. This is a separate bet of up to half their initial wager and pays 2 to 1 if the dealer has a natural. Of course, if the dealer does have 21, the main hand loses (unless the player also has a natural), and this is the reason for the name—the player is "insuring" the main hand against a dealer 21.

- Some casinos offer a *surrender* option, in which a player may elect not to play out his hand and forfeits only 50% of his initial wager. This might be something worth considering when the chances of beating the dealer are small—for example, when the player's hand is 16 and the dealer's upcard is a 10. Surrender comes in two versions: *early surrender*, where the option is available before the dealer checks his or her hole card for a possible natural, and the far more common and less player-friendly *late surrender*, which is only offered after the check for a natural is complete.

In light of these options, the outcomes of a $1 blackjack bet can be more than just "win $1" and "lose $1" that follow from a $1 bet on red at roulette, and this range of options leads to more complicated—and more interesting—mathematics. If the player doubles down, it's possible to win or lose $2, or to break even. A natural results in a win of $1.50. Surrendering introduces the possibility of a loss of $.50. If an initial pair is split, then it's possible to win

$2, win $1, break even, lose $1, or lose $2. If pairs can be resplit, or if a player can double down after splitting a pair, the list of possible outcomes grows.

Players may play more than one hand at most blackjack tables, provided that space is available. Most casinos, however, require that gamblers playing multiple hands bet more than the table minimum on the extra hands, frequently double the minimum on a second hand and triple the minimum on the third hand. While playing multiple hands gives no advantage in winning any particular hand, this can be a way to bet more money than the table maximum and thus win more if the dealer busts. If the hands are dealt face down, as is usually the case in one-deck and two-deck games, players are not allowed to look at the cards in one hand until any previous hands have been played out to completion.

Depending on the set of rules, the number of decks used, and the payoffs on naturals, the house advantage on a hand of blackjack, in practice, can range from essentially 0% to upwards of 4% for a player using basic strategy. (Deviations from basic strategy increase the HA.) At the outset, let us assume that we are playing single-deck blackjack under what are called "standard Las Vegas Strip rules" [137]. This name is used out of convenience; these rules are neither exclusive to the Strip nor universal thereon:

- Dealer stands on all 17s.
- Players may double on any two cards, and may double after splitting a pair.
- Aces may be resplit, though only 1 card is dealt to split aces.
- Naturals pay 3 to 2.

With these rules in force, single-deck blackjack is nearly an even game for a player using basic strategy. Table 5.1 on page 212 lists the effect on the house edge of various rules changes.

In the first few lines of Table 5.1, we see that the house edge increases as the number of decks in use goes up, but the effect levels out pretty quickly, and the difference between 6- and 8-deck games is only .03%—almost negligible to a basic strategy player. A card counter or other sort of blackjack advantage player might look for strategies to eliminate even this small a change in the HA. Whatever additional advantage a casino may gain from using 12 or more decks is outweighed by the inconvenience of handling and shuffling that many cards. Time spent shuffling is, after all, time not spent dealing hands and collecting the HA, on the average, from every hand dealt. A continuous shuffling machine (CSM) cuts down on shuffle time and allows more hands per hour by immediately recycling discards into the shoe and reshuffling them. With a CSM, every hand is effectively dealt from a fresh shoe.

A particularly onerous and common rule change reduces the payoff on a natural from the traditional 3–2 to 6–5. On a $10 bet, this changes the payoff on a natural from $15 to $12. While this may seem insignificant due to the scarcity of naturals, Table 5.1 shows that this rule raises the HA by

TABLE 5.1: Effects of rule changes on blackjack house advantage [13, 42, 47]. Negative values favor the player.

Rule change	Effect on HA
Deck Size	
Two decks	+0.32%
Four decks	+0.48%
Six decks	+0.54%
Eight decks	+0.57%
Continuous shuffling machine used	+0.30%
Double Downs	
No double downs on soft hands	+0.13%
Double down only on 10 or 11	+0.26%
No double downs on 9	+0.13%
No double downs on 10	+0.52%
Double downs only allowed on 11	+0.78%
No double downs allowed	+1.60%
Double down on two or more cards	−0.24%
Double down after pair splitting	−0.14%
Pair splitting	
No non-ace pair resplitting	+0.03%
No splitting of aces	+0.18%
Pair splitting forbidden	+0.40%
Resplitting aces allowed	−0.06%
Draw more than one card to split aces	−0.14%
Payoffs	
Dealer wins ties	+9.00%
Naturals pay 6 to 5	+1.39%
Naturals pay even money	+2.32%
Naturals pay 2 to 1	−2.32%
Six cards under 21 automatically wins	−0.15%
Soft 17s	
Dealer hits soft 17	+0.20%
Surrender	
Late surrender	−0.06%
Early surrender	−0.62%

1.39%—more than a threefold addition to the HA in a 1-deck game. This rule change finds its origins in Super Fun Blackjack (page 216). While 6–5 payoffs on naturals began with Super Fun 21, the rule change found fast approval at many casinos, especially on blackjack tables with lower bet minimums.

These percentages may be combined to determine the house edge for a blackjack game with a specified set of rules. Figure 5.2 shows the rules screen of a single-deck electronic blackjack game on offer at a downtown Las Vegas casino.

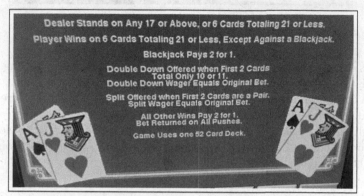

FIGURE 5.2: Electronic blackjack rules.

The following game parameters affect the HA:

- Blackjacks pay even money ("2 for 1"): HA +2.32%.
- Doubling down only permitted on 10 or 11: HA +.26%.
- No pair resplitting: HA +.03%.
- Six cards under 21 automatically wins: HA −.15%.

  These last two rules are included in part because of the difficulty of displaying three or more player hands, or a hand of 7 or more cards, on the video screen.

Adding up the cumulative effects of these rule changes results in a game with a house edge of 2.46%.

At the same downtown casino, a neighboring blackjack machine used the rules shown in Figure 5.3, with no option to split pairs or double down.

The rules for this game affect the HA as follows:

- Blackjacks pay even money: HA +2.32%.
- No pair splitting: HA +.40%.
- No opportunity to double down: HA +1.60%.
- Six cards under 21 automatically wins: HA −.15%.

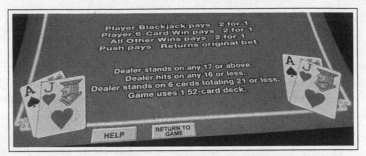

FIGURE 5.3: Alternate electronic blackjack rules.

This game—which gave players the option of defining their betting unit as $100—is much less friendly to players, with a casino edge of 4.17%.

The two games discussed here appear identical until the rules screens are revealed. As with video poker, it's essential to review the rules in force before settling in for a session of electronic blackjack. in particular, many electronic blackjack machines pay only even money on naturals, which has a big effect on the house edge.

Side Bets

Many new ideas and game variations have been proposed for blackjack bets. In part this is because blackjack is in part a game where skill at making choices plays a role, and so if there's a new bet offering a chance for players to make incorrect choices that favor the casino, that bet may seem attractive to some casino officials. In part this may be because blackjack using basic strategy is nearly an even game, and these side bets offer the casino an opportunity to increase its profits, as these bets frequently have a high house edge.

- The *Royal Match* bet allows players to make a side bet on their first two cards. The pay tables for Royal Match vary somewhat; one version pays 2½ to 1 if the first two cards are suited (the "Easy Match") and 25 to 1 if they are the king and queen of the same suit (the "Royal Match"). An additional option is to offer a progressive jackpot for the "Crown Treasure," which happens when both the player and dealer have a Royal Match.

- *Instant 18* is a separate side bet that must be made alongside a regular blackjack wager. When you make an Instant 18 bet, you are effectively playing a second hand that is assigned a value of 18. No cards are dealt to this phantom hand.

- One hallmark of a good blackjack side bet is its ability to be resolved before players start drawing cards, and *Over/Under*, which was launched at Caesars Tahoe in Stateline, Nevada (now Bally's Montbleu) in 1988, fits that description [128]. The player may wager that his or her initial 2-card hand will be either over 13 or under 13, with aces counted as 1 for the purpose of this bet. On a dealt 13, both bets lose. Over/Under does not involve the dealer's upcard, which makes it easier to administer. Once the wager is settled, the blackjack hands are played out against the dealer as usual.

- *Jack-A-Tack* is a side bet on offer in 2021 at Circus Circus Casino in Las Vegas. The bet is resolved on the player's initial 2 cards, with payoffs based on the number and type of jacks present. The highest payoff incorporates the dealer's first 2 cards as well. Table 5.2 shows the payoffs offered at Circus Circus. One-eyed jacks are the $J\spadesuit$ and $J\heartsuit$: the cards where the jacks appear in profile.

TABLE 5.2: Jack-A-Tack pay table

Hand	Payoff odds
4 one-eyed jacks (player and dealer)	299–1
2 one-eyed jacks (player)	100–1
2 suited jacks	100–1
Any 2 jacks	25–1
Total of 20	5–1

As is standard with side bets, only the highest winning combination is paid; a player hand of $J\clubsuit\ J\clubsuit$ pays only the 100–1 offered for 2 suited jacks, not the other payoffs lower in Table 5.2. If the player holds 2 one-eyed jacks and the dealer's upcard is also a one-eyed jack, then this bet will have to wait until the hand is fully played out for the dealer's hole card to be revealed and the payoff determined to be 100–1 or 299–1. This sets Jack-A-Tack apart from a bet like Over/Under that is resolved before any additional cards are dealt.

Game Variations

- Bob Stupak, whom we first met in Example 4.2.1 in connection with Crapless Craps, also invented a variation of blackjack called *Double Exposure 21*, in which both of the dealer's cards are revealed to the player. This game variation was originally suggested by gaming mathematician Richard Epstein under the name *Zweikartenspiel* in 1977 [22].

Revelaing both dealer cards, of course, gives the player a large advantage, but that is more than compensated for by new rules [132, p. 93]:

  - The house wins all ties, except ties between naturals, which the player wins.

This may lead to some unusual decisions that would never be made in ordinary blackjack, as when a player holding 20 against a dealer 20 must take a hit and hope for an ace, for 20 vs. dealer 20 is a losing hand.

– Naturals pay only even money, except naturals consisting of A♠ and J♠, which pay 2 to 1. A three-card 21 consisting of a suited 6, 7, and 8 also pays 2 to 1.

– Doubling down is only permitted on 9, 10, and 11.

– Pairs may be split only once.

Double Exposure II, introduced somewhat later, exposed the dealer's hole card only when it was a ten-count card, and forbade doubling and splitting pairs. The payoff on a suited 6-7-8 in Double Exposure II was triple instead of double.

- *Super Fun 21* claims to be "the most exciting way to play blackjack" according to the Orleans Casino's rack card (a promotional card that describes the rules and payoffs for a casino game) for the game. This version of blackjack is a 1999 single-deck variation on blackjack invented by Howard Grossman and designed in part as a defense against card counters. In Super Fun 21, players have the following additional options:

  – Double down on two *or more* cards, including after hitting or splitting pairs. If you have a three-card total of 11, you may double down—this is not allowed under standard blackjack rules.

  – Pairs, including aces, may be resplit up to three times, resulting in as many as four hands in play at a time.

  – Hands totaling less than 21 may be surrendered at any point, including after drawing one or more cards, doubling down, or splitting pairs, if the dealer does not have a natural.

  – A hand totaling 20 or less with six or more cards pays even money regardless of the dealer's hand, although if the hand has been doubled, the doubled amount is not an instant winner. (It does remain in play and is paid if the hand beats the dealer.)

  – A 21 with five or more cards pays 2 to 1 instantly, with the same restriction as for six-card 20s.

  – Player naturals are instant winners even if the dealer also has a natural. Naturals with both cards diamonds pay 2 to 1.

The following rule acts against the player and more than compensates the casino for the more player-friendly rules above:

  – Blackjacks other than in diamonds pay even money instead of 3 to 2 or 6 to 5.

- *Spanish 21*, introduced in 1995, is a blackjack variant played with six to eight decks of 48 cards from which all four 10s have been removed—this is sometimes called a "Spanish deck"[130]. Removing all of the 10s raises the casino's advantage by 2%, ½% per removed card, but a number of rule changes are added to make up for this effect [154]. There is some variability from casino to casino, but the alternate rules usually include these:

  - Players' naturals always win, even if the dealer also has a blackjack. Blackjacks pay 3 to 2.
  - All player totals of 21 beat dealer 21s. Other ties are pushes, as is the case in regular blackjack.
  - Players may double down on any number of cards, not just the first two.
  - Players may double down after splitting pairs.
  - Players may resplit pairs, including aces, for up to four hands.
  - A number of bonus payoffs for certain player hands are offered—for example, a five-card 21 pays 3 to 2, a six-card 21 pays 2 to 1, and a 21 consisting of seven or more cards pays 3 to 1.

- *Blackjack Switch* calls for each player to play two hands, with separate wagers of the same size on each. After the cards are dealt and the dealer has checked for a natural, players then have the option to switch the second cards in each hand from one hand to the other in order to create more favorable totals.

  Example 5.1.2. If a Blackjack Switch player is dealt $K\clubsuit$ $5\diamondsuit$ on one hand and $6\spadesuit$ $Q\spadesuit$ on the other, he or she may rearrange the cards to $K\clubsuit$ $Q\spadesuit$ and $6\spadesuit$ $5\diamondsuit$, thus changing two mediocre hands of 15 and 16 to a strong 20 and an 11. The player may then double down on the 11. ∎

The dealer plays only one hand and thus does not have the advantage given by switching cards. The player advantage is tempered by the following two rule changes:

  - All blackjacks pay even money rather than 3-2 or 6-5. Typically, a switched Ace-10 counts as a 21 rather than a blackjack, and so is not paid off immediately because it may be tied later by a dealer's 21.
  - If the dealer's hand totals a hard 22, then all remaining bets, including any 21's arising from switching cards, push. This rule turns a losing dealer hand into an automatic tie. A player 22 remains a losing hand, even if it's later tied by a dealer hard 22. (A dealer's soft 22 would be revalued at 12, and additional cards drawn.) This rule adds 6.91% to the HA, which more than compensates the casino for allowing card switching [114].

Other rules of blackjack—doubling down, splitting pairs, and insurance—remain as in standard blackjack.

5.2 Mathematics of Blackjack

Some of the mathematical questions associated with blackjack can be assessed with the mathematics we have already developed, without concern for the lack of independent events or the need to account for player decisions.

Example 5.2.1. Let's begin with a very simple question: What is the probability of being dealt a natural?

We consider only the player's hand, not the dealer's, and for simplicity, we will assume that we're dealing from a full single deck. Since a natural consists of a ten-count card and an ace, and since the order in which the cards arrive does not matter, we can consider each order separately. The probability of being dealt a ten-count card is $\frac{16}{52}$ and the probability of being dealt an ace is $\frac{4}{52}$. As cards are dealt, of course, these probabilities change, but since we're assuming that we start with a full deck, the only change is in the denominator, from 52 to 51. We then can compute

$$P(\text{Natural}) = \frac{16}{52} \cdot \frac{4}{51} + \frac{4}{52} \cdot \frac{16}{51} = \frac{32}{663} \approx .0482 \approx \frac{1}{21}$$

—so roughly 1 out of every 21 dealt hands will be a natural. In the calculation above, the first term in the sum covers the case where the ten-count card is dealt first, and the second covers the case where the ace is dealt first. Since these two events are mutually exclusive, the First Addition Rule may be used to add the probabilities.

If the hand is being dealt from the top of a six-deck shoe, a similar analysis gives

$$P(\text{Natural}) = \frac{96}{312} \cdot \frac{24}{311} + \frac{24}{312} \cdot \frac{96}{311} = \frac{4608}{97,032} \approx .0475,$$

slightly less than for a single-deck game. ∎

That having been said, it must be noted that this probability is a long-term average value. If, in a one-deck game, all four aces are dealt on the first round (see Exercise 5.13), no further naturals are possible until the deck is reshuffled, but if the first round contains no aces or ten-count cards, the likelihood of a natural in the second round increases. Deck composition and the dependence of each hand on previously dealt hands are what make blackjack mathematics tricky and explain why card counting works to improve the player's chance of winning big bets.

An alternate way to compute long-term average probabilities is the *infinite deck approximation*. As the name implies, the method assumes that blackjack is being dealt from a shoe containing infinitely many decks of cards. This simplifying assumption means that the probability of drawing any particular card, or type of card, from the shoe is constant, and independent of how many cards have been drawn previously. Under this approximation, the probability of drawing a natural is

$$P(\text{Natural}) = \frac{16}{52} \cdot \frac{4}{52} + \frac{4}{52} \cdot \frac{16}{52} = \frac{8}{169} \approx .0473 \approx \frac{1}{21\frac{1}{8}},$$

a value which is quite close to the value of .0482 calculated above for a full single deck.

Example 5.2.2. Given a two-card blackjack hand, the dealer's upcard, and the specific game rules in force, it is possible through repeated simulation to obtain an accurate estimate of the probability p of winning the hand. At what probability is surrendering the correct choice?

If you surrender, your expectation on a $1 bet is a flat –$.50. If you choose instead to play out the hand, your expectation is

$$E = (1) \cdot p + (-1) \cdot (1 - p) = 2p - 1.$$

This will be greater than –$.50 if $p > .25$, so if you have at least a 25% chance of winning, surrendering is not the optimal play. Turning this around, if your hand has greater than a 75% chance of losing, you should surrender if the option is available. ∎

The challenge here is that p is not a probability that is readily calculable from the three visible cards. Blackjack experts have, through the repeated simulation mentioned above, identified certain hands that should be surrendered, and these are listed in Table 5.3.

TABLE 5.3: Strategy for surrender [97]

Dealer upcard	Surrender
Ace	10/6
T	T/6, T/5, T/7, 9/6, 7/7

Note that the composition of the hand matters as well as the total: no soft hands should be surrendered, and while you should surrender T/6 against a dealer ace, you should not surrender 9/7 or 8/8, which also add up to 16. A pair of 8s should be split rather than surrendered, even against a dealer ace.

Example 5.2.3. Should you make the insurance bet?

If the dealer's upcard is an ace, you will be offered the opportunity to make what is effectively a side bet that the dealer has a natural. We assume that we're looking at the first hand of a new single deck, and that the only three cards known to you are the two in your hand and the dealer's ace. If you have no ten-count cards in your hand, the probability of a dealer natural is $\frac{16}{49}$. Since the insurance bet pays 2 to 1 and is limited to half your main bet, we have the following expected value for a $1 main bet:

$$E = (1) \cdot \frac{16}{49} + (-.50) \cdot \frac{33}{49} = -\$.0102.$$

The HA on a 50¢ bet is thus 2.04%—not awful, but this assumes that you hold no tens. If you have one or two, the dealer has a smaller chance of completing a natural, and thus the HA of an insurance bet increases.

Holding one ten, we have

$$E = (1) \cdot \frac{15}{49} + (-.50) \cdot \frac{34}{49} = -\$.0408,$$

and the HA on the insurance bet is 9.16%.

If you hold two tens, then the expectation drops further, to

$$E = (1) \cdot \frac{14}{49} + (-.50) \cdot \frac{35}{49} = -\$.0714$$

—an HA of 14.28%. ∎

But what if you hold a natural yourself? Some casinos will offer even money on a player natural against a dealer ace—offering to settle the main bet as though an insurance bet had been made without the need for the player to make the bet. This even money option arises from the fact that if the dealer has a natural, your main bet pushes but a hypothetical insurance bet of half the main wager pays off at 2 to 1. The net effect is a profit of the amount originally wagered.

Psychologically, there is some appeal to this bet—at least you win even money on your one-hand-in-21 natural, even if the dealer ties you. You forfeit the extra 50% payoff at a 3 to 2 table; that's equivalent to making an insurance bet and losing it. But what does the mathematics say?

As we might expect, the casinos aren't offering the even money option in an effort to be nice to players. There's something in it for them—let's do the math. There are 15 tens left in a 49-card deck. By taking even money, your return, regardless of the dealer's hand, is exactly $1. If you decline even money, which is equivalent to not making an insurance bet, you win $1.50 if the dealer does not have a natural and push if he does, so your expectation is

$$E = (1.50) \cdot \frac{34}{49} + (0) \cdot \frac{15}{49} \approx \$1.0408,$$

and declining even money is a better call by about 4.08%.

If, through bad fortune or poor judgment, you are playing blackjack at a casino that only pays 6 to 5 on naturals, then the expectation if you decline even money and the implied insurance bet is

$$E = (1.20) \cdot \frac{34}{49} + (0) \cdot \frac{15}{49} \approx \$.83,$$

so if the casino should offer even money ($E = \$1$) for your natural, take it. Of course, since casino management knows this, it is unlikely that you will even be offered even money on a natural if the casino pays only 6 to 5.

With an eye on the "all tie on dealer 22" rule in Blackjack Switch, it should be noted that 12 is the second most common two-card count, trailing only 20. This is in part because any first card can result in a two-card 12, whereas if your first card is a 9, there is no way to have a two-card total of 8. The probability of a two-card 12 in a single-deck game is

$$p = 2 \cdot \frac{4}{52} \cdot \frac{16}{51} + 2 \cdot \frac{4}{52} \cdot \frac{3}{51} + \frac{24}{52} \cdot \frac{4}{51} = \frac{248}{2652} \approx .0935,$$

or about once every 10.69 hands.

In this equation, the first term considers the case where the first card is a 2 or a ten-count card, the second covers the case where the dealt 12 is a pair of aces or a pair of 6s, and the third term collects all other cases leading to a hand of 12.

This calculation includes the hands 6-6 and A-A, which the player should often split rather than play out as 12s. When considering the possibility of a dealer drawing out to a total of 22, of course, these hands will not be split, for the dealer may not split pairs.

If we consider this question using the infinite deck approximation, then the probability of a two-card 12 is

$$p = 2 \cdot \frac{4 \cdot 16}{52 \cdot 52} + 2 \cdot \frac{4}{52} \cdot \frac{3}{52} + \frac{24}{52} \cdot \frac{4}{52} = \frac{31}{338} \approx .0917,$$

a respectably close approximation.

Example 5.2.4. How does the "different payoff on naturals" rule in the alternate blackjack games affect the player's income from naturals?

If we use the result of Example 5.2.1, we can conclude that once in every 21 hands, the player will receive a natural. If we're playing standard blackjack with a 3 to 2 payout, the extra money will be 50¢ per \$1 bet per 21 hands, assuming *flat*, or constant, bets. A 6 to 5 payout gives the player an extra 20¢ per \$1 bet per 21 hands, which is a 60% decrease from a 3 to 2 payoff game.

In either Super Fun 21 or Double Exposure 21, there is one type of natural that will pay an extra \$1 per dollar bet. The probability of a natural in diamonds (as in Super Fun 21) is

$$2 \cdot \frac{1}{52} \cdot \frac{4}{51} = \frac{2}{663} \approx .003,$$

so the extra money accruing to the player is about $1 per $1 bet per 331.5 hands, or 3.17¢ per $1 bet per 21 hands. This is less than the increased expectation of regular blackjack by a factor of about 16.

For Double Exposure 21, the probability of receiving the one blackjack that triggers a 2 to 1 payoff is

$$2 \cdot \frac{1}{52} \cdot \frac{1}{51} = \frac{1}{1326} \approx .00075,$$

so a player can expect to win an extra $1 per $1 bet per 1326 hands, or .016¢ per $1 bet per 21 hands, about $\frac{1}{63}$ of the payoff of standard 3 to 2 blackjack. ∎

Blackjack Switch lacks even this single exception and so eliminates all of the advantage of a natural. While the option to switch cards means that Blackjack Switch leads to more two-card 21s than the other games, 21s arising from a card switch aren't considered natural 21s—not that that matters for the payoff.

We cannot fail to see the true impact of this rule change: since so much is being taken from the players with the different payoff on naturals, the casinos can afford to give quite a bit back with the new rules, especially considering that players might not adjust their playing strategy to account for these rules and thus hand over even more percentage points of advantage to the house. In light of this calculation, we are not surprised to see that switching from 3 to 2 to 6 to 5 payoffs on naturals, as many casinos in Las Vegas have done on their lower-limit or single-deck games, produces a larger HA.

That having been said, it is worth noting that rules can be changed. Some casinos have offered Double Exposure blackjack with a 3-2 payoff on naturals, and this payoff turns a house edge into a positive player expectation ranging from .4% to 2.1%, depending on the other rules in force [149].

On page 214, we introduced the Royal Match wager, which is an optional side bet on a player's first two cards whose resolution does not affect the play of the hand. Consider the Royal Match bet that pays 2½ to 1 if the first two cards are suited (the "Easy Match"), 25 to 1 if they are the king and queen of the same suit (the "Royal Match"), and a progressive jackpot for the "Crown Treasure," which happens when both the player and dealer have a Royal Match. Leaving out the Crown Treasure progressive bet for the moment, is this a bet worth making?

Once again, we begin by assuming a full single deck. For the Easy Match bet, the first card can be any card at all; we are simply looking for the probability that the second card is the same suit as the first.

$$P(\text{Easy Match}) = \left(\frac{52}{52}\right) \cdot \left(\frac{12}{51}\right) = \frac{12}{51} = \frac{4}{17} \approx .2353.$$

It should be noted that this calculation includes the probability of drawing a Royal Match. We will account for that when we develop the probability distribution.

For the Royal Match payoff, two things must happen:

1. The first card dealt to the player must be a king or queen.

2. The second card must be the rank not dealt in step 1, *of the same suit.*

It follows that

$$P(\text{Royal Match}) = \frac{8}{52} \cdot \frac{1}{51} = \frac{2}{663} \approx .0030.$$

We find that the probability of hitting the Easy Match but *not* the Royal Match is $.2353 - .0030 = .2323$. Defining X to be the return on a \$1 Royal Match bet gives the following probability distribution:

x	2.5	25	-1
$P(X = x)$	.2323	.0030	.7647

—so this is a bet you will lose approximately $\frac{3}{4}$ of the time.

For a \$1 bet on the Royal Match, we have

$$E = (2.5) \cdot .2323 + (25) \cdot .0030 + (-1) \cdot .7647 = -\$.10895,$$

and the HA is about 10.9%.

In the infinite deck approximation, we find that

$$P(\text{Easy Match}) = \frac{52}{52} \cdot \frac{12}{52} = \frac{12}{52} = \frac{3}{13} \approx .2308,$$

and that

$$P(\text{Royal Match}) = \frac{8}{52} \cdot \frac{1}{52} = \frac{1}{338} \approx .0030.$$

The difference between this last probability and the one calculated in the single-deck case is

$$\frac{2}{663} - \frac{1}{338} = \frac{1}{17,238} \approx 5.80 \times 10^{-5},$$

so little is lost in assuming an infinite shoe for the Royal Match payoff.

If the Crown Treasure progressive bet is active, then the expected value of the Royal Match bet depends on the size of the progressive jackpot, and there will be an amount past which this bet favors the player. Let us denote the amount of the jackpot by J. The probability of winning the Crown Treasure bet is the probability that player and dealer both receive a Royal Match. For the dealer's hand, under the assumption that the player has a Royal Match, the probability of a second Royal Match is

$$P = \frac{6}{50} \cdot \frac{1}{49} = \frac{3}{1225} \approx .0024,$$

and so the probability of a Crown Treasure is

$$p = \frac{2}{663} \cdot \frac{3}{1225} = \frac{1}{135,362.5} \approx 7.388 \times 10^{-6}.$$

This probability, insignificant though it may be, must be subtracted from P(Royal Match) above in computing the new expected value. We have

$$E = (2.5) \cdot .2323 + (25) \cdot (.0030 - p) + (J) \cdot p + (-1) \cdot .7647,$$

or

$$E = -.10895 + p \cdot (J - 25).$$

If we set this last quantity equal to 0 and solve for J, we get

$$J = \frac{.10895}{p} + 25,$$

which yields J = 14,772.74438, so *if* the progressive jackpot for the Crown Treasure exceeds \$14,745.74, Royal Match is a bet that favors the player, since the expectation above will then be greater than zero.

We return now to the Over/Under bet. Table 5.4 shows the distribution of the sum of a player's 2 cards using the infinite deck approximation.

TABLE 5.4: Over/Under: Sum of a player's first 2 cards. Aces count as 1, infinite deck approximation in force.

Hand	Probability	Hand	Probability
2	.0059	14	.0769
3	.0118	15	.0710
4	.0178	16	.0651
5	.0237	17	.0592
6	.0296	18	.0533
7	.0355	19	.0473
8	.0414	20	.0947
9	.0473		
10	.0533		
11	.0947		
12	.0888		
Under:	.4497	**Over:**	.4675
Probability of 13: .0828			

Using this information, we can compute the house advantages for Under and Over.

$$E(\text{Under}) = (1) \cdot .4497 + (-1) \cdot .5503 = -.1006,$$

for a HA of 10.06%, and

$$E(\text{Over}) = (1) \cdot .4675 + (-1) \cdot .5325 = -.0650,$$

giving a HA of 6.50%.

Example 5.2.5. A challenge in developing blackjack side bets is separating the side bet from the play of the main game to minimize confusion, and so many such bets are resolved before players begin drawing cards. The *Lucky Lucky* side bet offers a range of payoffs based on the player's cards combined with the dealer's upcard, and pays off on any three-card total of 19 through 21, with a top prize of 200 to 1 if the three cards are all 7s of the same suit.

The probability of this last payoff depends on the number of decks in play (don't make this bet if it's offered at a double-deck game!). For a six-deck game, we are interested in the probability of drawing three cards and having them all be the same 7. Starting from a full shoe gives

$$p = \frac{4 \cdot \binom{6}{3}}{\binom{312}{3}} = \frac{80}{5,013,320} \approx 1.596 \times 10^{-5}.$$

200 to 1 scarcely seems adequate payoff for the incredibly long odds you've beaten in pulling this hand. ∎

An advantage to the *Bonus Blackjack* side bet is that it is only available on the first hand after a shuffle and cut, and so can be easily assessed without the need to account for changing deck composition. An individual player may bet on their own hand, the dealer's hand, or both. The bet pays off at 15–1 if the chosen hand gets a natural. If a player bets both hands and their hand is dealt A♠ J♠, a progressive jackpot is awarded.

In a 6-deck shoe, the probability of a natural to either hand is

$$2 \cdot \frac{24}{312} \cdot 96311 = \frac{192}{4043} \approx .0475,$$

considerably less than $\frac{1}{15}$. The HA of either the Player or Dealer bet is 24.02%.

The probability of a player A♠ J♠ is

$$2 \cdot \frac{6}{312} \cdot \frac{6}{311} = \frac{3}{4043}.$$

The expected value of this $2 bet is a function of the progressive jackpot J, and is positive if

$$E(J) = (J) \cdot \frac{3}{4043} + (-2) \cdot \frac{4040}{4043} > 0,$$

which happens if the jackpot is at least $2693.33.

5.3   Basic Strategy

Further calculation of blackjack probabilities will assume that the player is using the correct basic strategy, so a discussion of what that means is in order here.

Blackjack is a game that offers players choices in how to play their hands, and so it is reasonable to conclude that some choices are better than others in light of the information available: the player's cards and the dealer's upcard. Some of these choices are very easy: don't hit a hard 19 regardless of what the dealer is showing, for example. Other choices are more subtle: if you hold hard 16 and the dealer is showing a queen, is it better to hit or to stand? Certainly a hand of 16 will lose if the dealer does not bust, but starting with a queen means that the dealer has a strong start and a better probability of holding or drawing to a 17 or higher. Hitting, though it runs a high probability of busting, may be worth the risk if the hand needs to improve in order to win.

Prior to the formal development of what we now call basic strategy, some gamblers had begun to identify certain strategy plays that seemed to be better than others. Sidney H. Radner devised a "Calculation Method" that suggested, with a simple plausibility argument but without formal mathematical justification, that the choices listed in Table 5.5 were the best for the player.

TABLE 5.5: Radner's Calculation Method for blackjack [94]

Always stand on 16 or higher.
Hit 15 only against a dealer 7 or higher.
Always split aces and 8s.
Never split 10s, 9s, 7s, 6s, 5s, or 4s.
Double down on 9, 10, or 11, but only against a dealer 2–6.
Only take insurance when playing a betting system.

It is noteworthy that Radner's recommendations did not distinguish between soft and hard hands. Some of these decisions have been mathematically verified as optimal strategy, while others have been superseded by more thorough mathematical analysis.

The formal history of what we now call basic strategy begins with "The Optimum Strategy in Blackjack," a paper by Roger Baldwin, Wilbert Cantey, Herbert Maisel, and James McDermott that was published in the September 1956 issue of the *Journal of the American Statistical Association* [5]. The four authors were mathematicians serving in the Army at the Aberdeen Proving Grounds in Maryland, and their casual blackjack games inspired the foursome to analyze the game mathematically. This paper was the first to suggest that a set of rules for player action, which has now evolved into basic strategy, would lead to the best possible return for players. For the first time, the idea that the player's position in blackjack could be improved by skillful play was backed up with a rigorous mathematical argument. Among other revelations, this paper was the first to advocate, with mathematical justification, that a pair of 8s should always be split and that standing on a 12 against a dealer 4, 5, or 6 is the better play.

Given the player's cards and the dealer's upcard, basic strategy gives the best choices for the player: when to hit, stand, double down, or split pairs for the best long-term results. The key word here is *long-term*: basic strategy, for example, tells a player to hit rather than stand when holding 15 against a dealer 8. Note that the only way a player 15 can win is if the dealer busts, since a completed dealer hand will always be 17 or greater. While hitting a 15 will certainly bust some hands—more often than not, in fact—computer simulation indicates that this action is best for the player, in the sense that, over a lifetime of gambling, you will win more (or lose less, which is effectively the same thing) by hitting and taking the chance of busting than by standing and hoping the dealer will bust.

The probabilities derived in this manner which give rise to basic strategy are often experimental probabilities, this being one case where it's easier to generate large numbers of simulated hands and draw conclusions from data than to account theoretically for the effects of player decisions on game outcomes.

Example 5.3.1. What does this mean mathematically? If the dealer's upcard is an 8, the probability distribution for the final value x of the hand is (from [47]):

x	17	18	19	20	21	> 21
$P(x)$	.131	.363	.129	.068	.070	.239

so by standing on 15, you have only a 23.9% chance of winning; you will lose roughly three of every four hands. By hitting your 15, you take the chance of busting ($P(\text{Bust}) \approx \frac{7}{13} \approx .538$), but you also have about a 46% chance of improving your hand and possibly beating a nonbusting dealer hand. Your actual chance of winning if you hit is not quite 46%, but it's better than 23.9%, which is why basic strategy directs a player to hit 15 against an 8. ∎

The goal of basic strategy is simple: to get the most money in play, via double downs and splitting pairs, when conditions favor the player, and to minimize exposure by forgoing these options when conditions are less favorable. If the dealer's upcard indicates that his hand is likely to bust—say, a 4, 5, or 6—then we double down and split pairs more frequently, and do not risk busting ourselves. When the hand is strong—a ten or ace showing, for example—then we take more risks as we try to compete against a hand that is more likely to be close to 21, and we don't invest additional money under the less favorable circumstances.

Table 5.6 contains the complete basic strategy for a multideck game. A simplified strategy, suitable for beginners, can be stated in a few straightforward rules:

1. **Always split 8s and aces.** Sixteen is a very weak hand, and turning it into two hands starting at 8 gives you a fighting chance against the dealer. You may not win both hands, but your chance of breaking even, by winning one of the two new hands, increases. A player hand starting with an ace has a 52% edge over the casino; splitting a pair of aces gives you two such hands.

   Aces and 8s illustrate the two types of pair-splitting: offensive and defensive. Splitting aces is an offense-minded move: the gambler has good cards, and by dividing the pair into 2 strong hands, is in a position to win more money. Splitting 8s is a defensive decision: by breaking a weak hand of 16 into two slightly better hands starting with an 8, the player is making the best of an admittedly poor starting hand.

2. **Never split 4s, 5s, or 10s.**

   - 8 may not be the best starting hand, but two hands starting with a 4 is a weaker situation. Hit this hand.
   - Splitting 5s means breaking up a hand of 10, which is a good total to draw to. Indeed, most of the time, a basic strategy player should double down on 5-5.
   - Splitting 10s can be tempting, since starting a hand with a 10 gives you a 13% advantage over the house, but it means turning your back on a very strong hand of 20. In addition to being an unwise play, splitting 10s will almost certainly increase casino scrutiny and get you labeled as either a bad player or a card counter.

3. **Always double down on 11, and double down on 10 except against a 10 or ace.** In the infinite deck approximation, holding an 11 gives you a $\frac{4}{13}$ chance of drawing one card and getting a unbeatable 21. If you hit a 10 with one card, you have that same $\frac{4}{13}$ chance of drawing a 10 and reaching 20, and if you draw an ace $\left(p = \frac{1}{13} \right)$, you have a 21. In either case, you have a strong hand, and it's time to get more money in play.

4. **Hit any soft hand with a total less than 19.** A soft hand gives you two chances at improving your holdings. If drawing a card puts your total over 21, revalue the ace as 1 and follow the basic or simplified strategy for your new hand.

5. **If the upcard is 2 to 6, stand on any hard count of 12 or higher.** In short, don't risk busting when facing a weak dealer upcard.

TABLE 5.6: Basic strategy for six-deck blackjack [47]

Player's Hand		Dealer's upcard									
		2	3	4	5	6	7	8	9	T	A
	17+	S	S	S	S	S	S	S	S	S	S
	16	S	S	S	S	S	H	H	H	H	H
	15	S	S	S	S	S	H	H	H	H	H
	14	S	S	S	S	S	H	H	H	H	H
No pair	13	S	S	S	S	S	H	H	H	H	H
or ace	12	H	H	S	S	S	H	H	H	H	H
	11	D	D	D	D	D	D	D	D	D	H
	10	D	D	D	D	D	D	D	D	H	H
	9	H	D	D	D	D	D	H	H	H	H
	8−	H	H	H	H	H	H	H	H	H	H
	A/9	S	S	S	S	S	S	S	S	S	S
	A/8	S	S	S	S	S	S	S	S	S	S
	A/7	S	DS	DS	DS	DS	S	S	H	H	H
Soft hand	A/6	H	D	D	D	D	H	H	H	H	H
(no pair)	A/5	H	H	D	D	D	H	H	H	H	H
	A/4	H	H	D	D	D	H	H	H	H	H
	A/3	H	H	H	D	D	H	H	H	H	H
	A/2	H	H	H	D	D	H	H	H	H	H
	A/A	SP	SP	SP	SP	SP	SP	SP	SP	SP	SP
	T/T	S	S	S	S	S	S	S	S	S	S
	9/9	SP	SP	SP	SP	SP	S	SP	SP	S	S
	8/8	SP	SP	SP	SP	SP	SP	SP	SP	SP	SP
	7/7	SP	SP	SP	SP	SP	SP	H	H	H	H
Pair	6/6	H	SP	SP	SP	SP	H	H	H	H	H
	5/5	D	D	D	D	D	D	D	D	H	H
	4/4	H	H	H	H	H	H	H	H	H	H
	3/3	H	H	SP	SP	SP	SP	H	H	H	H
	2/2	H	H	SP	SP	SP	SP	H	H	H	H
Never take insurance.											

Key:
D: Double down
DS: Double down if permitted, otherwise stand
H: Hit
S: Stand
SP: Split

6. **Against an upcard of 7 or better, stand only if your hand is 17 or higher.** The dealer is now more likely to have a strong hand, and you'll have to risk busting to improve your hand and have the best chance of winning.

7. **Never take insurance.** Example 5.2.3 gives the mathematical reason for this.

There are some places where this simplified strategy conflicts with basic strategy, but novices may find these rules easier to remember at first. Another option is to bring a wallet-size basic strategy card to the table—many casino gift shops sell such cards. There are also many blackjack Web sites (blackjack-info.com is one) that allow you to specify the parameters of the game and will then generate a basic strategy chart appropriate for that game.

For an examination of how basic strategy affects the house advantage, let us assume that we are playing six-deck blackjack under standard Las Vegas Strip rules (page 211). Suppose in addition that you choose to ignore basic strategy and simply adopt a "mimic the dealer" strategy by playing exactly as the dealer must: forgoing double downs and splits (since these are not available to the dealer), hitting on all 16s, and standing whenever your hand is 17 or higher, regardless of what card the dealer is showing. This strategy is rooted in the superficially plausible notion that since these are the rules that the dealer must follow, casinos must have devised them to give themselves the best advantage—so the savvy player should do the same [94].

Casinos will welcome you; indeed, they may compete for your business. In so doing, you are handing the house approximately a 5.46% advantage, more than they would get if you played American roulette [42]. Blackjack is symmetric under this approach, since both sides are using the same strategy. Remembering that the HA in blackjack derives from the double bust, and that the dealer's required strategy results in a bust 28% of the time, the HA starts at $(.28) \cdot (.28) = .0784$. We must reduce this to account for the 3 to 2 payoff on naturals available to the player but not to the dealer: this reduction is approximately $.5 \cdot \dfrac{1}{21} = 2.38\%$, since the player profits an additional half a bet every 21 hands on average (see Example 5.2.1). Subtracting gives an approximate HA of 5.46%.

Using the recommendations called for in the basic strategy table gives a player the benefits in Table 5.7, which have been determined by repeated simulation [42].

Adding these together shows that basic strategy can turn the HA from 5.5% to approximately 0%—a dead-even game. It is admittedly difficult to find a game played under these rules today: many casinos offering otherwise favorable games extract a price from players by paying only 6 to 5 on naturals unless the game has a very high minimum bet—$25 or more. This modification raises the HA by 1.39%, a change roughly equivalent to banning double downs. A 6 to 5 payoff on naturals is, by itself, good reason to seek out a different table,

TABLE 5.7: Basic strategy advantages

Option	Edge
Proper standing	3.2%
Doubling down	1.6%
Proper pair splitting	.4%
Hitting soft 17 & 18	.3%

or a different casino entirely, for your blackjack action. Moreover, casinos that require that a dealer hit soft 17s are giving themselves a chance to turn a weak dealer hand into a stronger one—admittedly at some risk—and increasing their advantage by about .20% [47].

Going in the other direction: From time to time, some casinos will offer a promotion or coupon where blackjacks pay 2 to 1. This adds 2.37% to the player's expectation in a six-deck game, turning a 1% HA into a 1.37% player edge. Benny Binion, longtime owner of the Horseshoe Casino (now Binion's) in downtown Las Vegas, used to offer this promotion at Christmas as a gift to his loyal players—but he limited his losses by restricting the bonus to bets of $5 or less and only allowing it on one hand per player at a time. If you were playing more than one hand at a time, only one hand qualified for the 2 to 1 payoff.

Of course, basic strategy errors do occur, whether because players find themselves in a rare situation that they have not memorized—for example, whether or not to split 9s against a 7 (you shouldn't, because your total of 18 will win if the dealer has a ten in the hole, which happens with probability $\frac{4}{13}$)—or because they are "feeling lucky" and acting on gut instinct rather than following basic strategy recommendations. Such errors, averaged across all players, are generally taken to add about .7% to the casino's advantage [153].

Variations

Blackjack variants—Double Exposure, Spanish 21, and others—require different strategies if the player is to play correctly and minimize the house advantage. In Double Exposure, for example, you must hit your hand if it is beaten in sight by the dealer's cards, even if—as when you hold 19 against a dealer 20—you would never draw a card in standard blackjack.

In Spanish 21, pursuing the bonuses for 21s with five or more cards sometimes calls for a player to hit hands with a number of small cards in hopes of catching the bonus. For example, a player would ordinarily stand if holding a hard 15 against a dealer 2 in standard blackjack. However, the following adjustments should be made to take full advantage of the bonus [93]:

- With 4 or more cards, such as 7332, hit hard 15 against a dealer 2.

- With 5 or more cards, hit hard 15 against a dealer 3 or 4.
- Finally, with 6 cards, hit hard 15 against a dealer 5 or 6.

The "5-Card Charlie" and "6-Card Charlie" are perhaps more common in casual home blackjack play than in the casino game. This rule variation, a version of which is incorporated into Super Fun 21 and some video blackjack games, states that a five- or six-card player hand totaling 21 or less is an automatic winner regardless of the dealer's hand.

To take full advantage of this rule change necessitates some small changes to basic strategy—to wit: occasionally drawing on some four- or five-card totals in hopes of catching a small fifth or sixth card and an immediate winning hand. For example, if you hold a soft 21 in five cards, such as *A*2224, drawing a sixth card puts you at no risk of busting and guarantees you a win, even if the dealer later draws out to the 21 that would tie you if you stood on your 21.

"Charlies" are perhaps best analyzed, once again, through simulation. Stanford Wong has described the necessary modifications to basic strategy if the 5-card Charlie payoff is active; this rule increases the player edge by 1.46%, *provided* that the player uses the correct modified basic strategy. Under the assumption that a 5-card Charlie pays even money, those modifications are detailed in Table 5.8. In some cases, these strategy changes call for forgoing doubling down on a soft hand so that additional cards may be drawn if the bonus is reachable.

TABLE 5.8: Basic strategy modifications for 5-card Charlie blackjack games [148]

4-card hands	Against a dealer
Hit soft 16	4
Hit soft 15	5
Hit soft 14	5 or 6
Hit soft 13	6
Hit hard 17	8, 9, 10, or ace
Hit hard 16	2 or 3
Hit hard 15	2 through 6
Hit hard 14	2 through 6
Hit hard 13	3 through 6
3-card hands	**Against a dealer**
Hit hard 13	2
Hit hard 12	4, 5, or 6
2-card hands	**Against a dealer**
Split aces	10 or ace
Split 3s	4, 5, or 7

A 6-card Charlie rule lowers the HA by .16%. Table 5.9 contains the recommended strategy changes for a game where any six-card hand is an automatic winner.

TABLE 5.9: Basic strategy modifications for 6-card Charlie blackjack games [42]

5-card hands	Against a dealer
Hit soft hands less than 20	Anything but 7 or 8
Hit hard 15 and less	2
Hit hard 14 and less	3 or 4
Hit hard 13 and less	5 or 6

Expert Blackjack Switch strategy comes in two parts. Both must be mastered if a player is to minimize the HA, which is .58% in a six-deck game. The first is a set of guidelines for how and when to switch cards between hands. Some decisions are simple: if you're dealt $Q\Diamond\ T\heartsuit$ and $K\heartsuit\ Q\clubsuit$, there is no point in switching, and switching in the two hands $9\Diamond\ 7\heartsuit$ and $4\heartsuit\ Q\Diamond$ turns a 16 and a 14 into the far better pair of a 19 and an 11.

However, what if your two hands are $9\Diamond\ A\spadesuit$ and $A\clubsuit\ 5\spadesuit$ against a dealer 7? Your soft 20 and soft 16 could be switched to a pair of aces—which could then be split—and a weak hard 14. It's not as clear what the better choice is. As with basic strategy, the question of when to switch hands can be answered by computer simulation of millions of hands. Each possible player hand can be assigned an expected value against a specified dealer upcard. Expected values range from 2.00 for a player natural against a dealer 2–9 to 0.43 for a hard 17 against a dealer ace, and the player should switch cards, or not, to get the highest possible total expected value from the 2 hands that result. Reflecting the facts that Blackjack Switch hands tie on a dealer 22 and that the most common hand totaling 22 is a deuce and 2 tens, all player hands except a natural have lower value against a dealer 2 than a dealer 3 [35].

In this example, the four possible hands have the following values:

Hand	Value vs. 7
$9\Diamond\ A\spadesuit$	1.71
$A\clubsuit\ 5\spadesuit$	0.95
$9\Diamond\ 5\spadesuit$	0.65
$A\clubsuit\ A\spadesuit$	1.34

Since the initial hands are worth 2.66 while the switched hands are worth 1.99, the better choice is not to switch cards [35].

Once the hands are switched or not, successful Blackjack Switch players need to modify standard basic strategy slightly to accommodate the "push on 22" rule [115]. These changes call for fewer double downs and splits, and are intended in part as protection against the high HA that "push on 22" affords the casino; this rule adds about 6.91% to the casino edge [114]. This is, in part, why a pair of aces has a lower expected value than a standard blackjack player might think. While proper Blackjack Switch strategy still calls for splitting all pairs of aces, a hand starting with an ace is slightly less valuable with "push on 22" in effect.

Some changes to blackjack basic strategy when playing Blackjack Switch are listed in Table 5.10.

TABLE 5.10: Selected basic strategy changes for Blackjack Switch [35]

Hit 5/5 against a T or A instead of doubling down
Hit 8/8 against a T or A instead of splitting
Hit hard 13 against a 2 instead of standing
Hit hard 9 against a 3, 4, 5, or 7 instead of doubling down
Hit hard 10 against a 9 instead of doubling down

Following both proper switching strategy and Blackjack Switch basic strategy keeps the HA at .45%. Switching correctly but using standard blackjack basic strategy nearly doubles that advantage, to .88%. Using the right basic strategy but never switching makes a huge difference: the HA is 9.12% [35].

5.4 Introduction to Card Counting

Basic strategy does not hold the answer to all blackjack questions. The following hypothetical blackjack situation was first described by mathematician Edward O. Thorp [135]:

> You are playing one-on-one against the dealer, and have determined that the last five cards remaining to be dealt are three 8s and two 7s. How should you play this hand?

The correct strategy is to bet as much as you can on the next hand. Borrow money if you have to, for you are a guaranteed winner.

You and the dealer will each be dealt a total of 14, 15, or 16, with one card remaining to be dealt. You stand—which you would not ordinarily do with a hard 14–16 against a dealer's 7 or 8 upcard—and the dealer must draw. If the dealer has 14, the remaining card must be an 8, and the dealer busts. If the dealer's total is 15 or 16, the remaining card—whether 7 or 8—is a certain bust card. Enjoy your win.

In practice, you would never see a game dealt down this far, nor would you know with 100% accuracy what the last five cards were. Even single-deck games reshuffle before getting down this far into the deck, and shoe games typically reshuffle with 1 to 1½ decks left. Bob Stupak once invented a blackjack variant called *Experto 21*, which used a single deck and was dealt down to the last card, but one card was burned at the start of the hand precisely to prevent endgame gambits such as this one. To counteract the obvious advantage this offered to a player with a good memory, naturals at Experto 21 paid only even money—which increases the HA by 2.32% [125].

Though the specific circumstances may be unlikely, Thorp's example points out the possible advantage to be gained from knowledge of the unplayed cards at blackjack. As we have repeatedly mentioned throughout this chapter, blackjack is different from other casino games in that successive trials are not independent. Each hand in blackjack depends in a meaningful way on the cards that have been dealt since the last shuffle, and as a result, probability calculations for blackjack must take this into account. The infinite deck approximation described in Section 5.2 assumes independence in its calculations and sacrifices a small amount of accuracy for ease of computation. Since we are looking specifically at how previous cards affect probabilities, we must discard the assumption of independence and thus the infinite deck simplification.

Thorp raised the profile of basic strategy as devised by Baldwin *et al.* and added the new twist of card counting with his 1962 book *Beat the Dealer* [134], in which he presented a mathematical argument describing how the composition of the undealt cards in a deck affected the probabilities of hands yet to come. In the ensuing years, blackjack players and mathematicians alike worked to determine an optimal strategy for keeping track of unplayed cards and using that information to adjust betting and playing strategies—and casino officials sought to counteract this new player advantage by changing the rules of the game and the procedures for its operation.

There is nothing mathematically complicated about card counting; all that is required for the most commonly used counting system is the ability to add and subtract 1, and to divide. (To do so under casino conditions is, as portrayed in the movie *21*, considerably more challenging.) It is not necessary, nor is it desirable, to try and track every card in a deck or shoe; all that most card-counting systems do is give the player a sense of the balance between high and low cards in the remaining undealt portion of the deck. This information, combined with the ability to adjust bet sizes and the willingness to learn exceptions to basic strategy, makes card counting a tool that can give a blackjack player an advantage over the casino.

Casino managers, of course, know this, and so the years since the publication of Thorp's book have seen a collection of changes to blackjack as casinos try to stop, or at least discourage, card counters. These changes include the introduction of more decks into the game, fewer hands dealt before reshuffling, and more conservative rules, as well as increased scrutiny of player activity, especially bet fluctuation.

Thorp's Five-Count

The simplest card-counting system described in *Beat The Dealer* involves tracking only the 5s. This is best suited for single- and double-deck games; when the book was first published, shoes containing four, six, or eight decks were nowhere near as common as they would become. Since the event that triggers a change in betting and playing strategy is the removal of all the 5s, this is far less viable in shoe games, where the likelihood of all the 5s being dealt out before the cards are shuffled is very small.

Why the focus on the 5s? Since the dealer must draw when holding 16, the absence of 5s renders the deck less favorable to the house. Thorp's calculations measured the effect on the house advantage of the removal of all the cards of each rank from a single deck, and his results are shown in Table 5.11. It can be seen that the biggest boost to the player comes when all of the 5s are removed, whereas removal of all of the aces or 9s increases the house edge.

TABLE 5.11: Effect of card removal on blackjack house edge [134]

Card Rank	A	2	3	4	5
% HA Change	2.42	−1.75	−2.14	−2.64	−3.58
Card Rank	6	7	8	9	T
% HA Change	−2.40	−2.05	−0.43	0.41	−1.62

It follows that, if all of the 5s have been dealt, and there are sufficient cards remaining in the deck to complete the next hand, the player has a considerable edge, and a larger bet is therefore called for. Some small changes in basic strategy are appropriate in a game with no 5s remaining; among these is that a pair of 10s should be split against a dealer's upcard of 6 [134].

The High-Low Count

A simple but powerful counting system, and arguably the most popular, is the *high-low count*. The idea behind the high-low count is that a deck with a surplus of high cards favors the player, in the sense that the probability of naturals and of drawing a high card on a double down both increase. Of course, the probability of the dealer pulling a natural also increases, but since player naturals are paid 3 to 2 while dealer naturals lose only the amount bet, this is still a benefit for the player.

Thorp's original high-low card-counting system simply tracked the ratio of non-10s to 10s in a deck, a ratio that starts at 36/16 in a full deck [134]. This was soon eclipsed by more sophisticated counting methods. In the high-low counting system, the cards are assigned the point values in Table 5.12.

A complete deck, or shoe of several decks, has initial count 0, since there are exactly as many high cards as low cards in a deck. Cards are counted as

TABLE 5.12: High-Low Count point values

Cards	Value
Low: 2, 3, 4, 5, 6	+1
Neutral: 7, 8, 9	0
High: T, J, Q, K, A	−1

they appear: add 1 to the count for every low card dealt and subtract 1 for every high card. The total is called the *running count*, RC.

Example 5.4.1. At the start of a four-player single-deck game, suppose that the cards dealt are J/9, K/5, K/J, and 2/Q, with a 4 for the dealer's upcard. The running count is $(-1 + 0) + (-1 + 1) + (-1 - 1) + (1 - 1) + 1 = -2$. ∎

For playing purposes, the running count must be converted into the *true count*, TC. This is done by dividing the RC by the number of decks remaining to be dealt. This divisor is estimated from the progress of the hand; a look at the discard rack on the table can help with the estimation. This modification gives a per-deck accounting of the excess or deficiency of high cards. A running count of +4, signifying an excess of four high cards over low cards, is a lot more meaningful if only one deck remains than if three are left in the shoe.

Example 5.4.2. Completing the hand above: Since the dealer's upcard is a 4, basic strategy dictates that all four players, who hold 19, 15, 20, and 12, stand. No further cards were drawn and the count remained at −2. The dealer's hole card was a 7 (RC still −2) and he drew a 10 (RC now −3) for a total of 21, beating all four players. Eleven cards have been dealt, leaving roughly $\frac{3}{4}$ of a deck—in general, the number of decks remaining is estimated to the nearest half- or quarter-deck. The TC is then $-3 \div (3/4) = -4$, meaning that low cards currently exceed high cards, making the game less favorable to players. Consequently, they should bet no more than the table minimum. ∎

There are two uses for the TC. The first is to indicate a point when the player's initial bet should be increased because the cards are favorable and a good hand is more likely, or conversely (as in the example above), when the cards favor the casino and bets should be kept as small as possible. To be precise, every increase of 1 in the true count translates into approximately a .5% advantage for the player [137]. Since the game starts out with a house edge of about .5% when using many common rule sets including a 3-2 payoff on naturals, a count of +1 makes the game essentially even, and if the true count is +4, players have a 1.5% edge over the casino.

The second purpose of the TC is to highlight times in the course of a deck or shoe when the composition of the remaining cards is such that certain deviations from basic strategy are, probabilistically, more favorable to the player.

Once again, these true counts, called *index numbers*, have been determined by computer simulation.

For example, consider the insurance bet. Since you're basically making a side bet that the dealer has a ten in the hole, a positive count indicating more high cards might indicate that that bet is worth making. Since insurance pays 2 to 1, this bet has a player advantage (independent of the player's hand) if the ratio of 10s to non-10s exceeds 1:2.

Suppose that on the first round of a one-deck game, 11 cards were dealt: 2 tens and 9 lower cards. On the second round, the dealer shows an ace and your hand is a pair of 5s. The composition of the deck is now $16 - 2 = 14$ tens cards and $36 - 12 = 24$ nontens. Since the ratio of tens to nontens is $\frac{14}{24} > \frac{1}{2}$, an insurance bet has an expectation in your favor. Specifically, the expected value of a 50¢ insurance bet is

$$E = (1) \cdot \frac{14}{24} + (-.50) \cdot \frac{10}{24} = \$\frac{9}{24} > 0,$$

so you should make the insurance bet.

This does not account for the fact that if the insurance bet wins, your original $1 bet loses unless you also have a natural, and a careful analysis of game options will have to take that into account.

Translating this into the world of card-counting, this means that if the TC is +3 or greater, the insurance bet favors the player. This index number takes into account the fact that aces as well as tens count −1 in the high-low count, hence the count by itself is not a perfect indicator of the ratio of tens to nontens remaining in the shoe. At the other end of the spectrum, a highly negative true count, indicating an abundance of low cards over high ones, may mean that hitting certain hard hands despite the advice of basic strategy—for example, hitting a hard 12 against a 5 at a TC of −2—is called for. Eighteen deviations from basic strategy for a four-deck game where the dealer stands on all 17s—dubbed the "Illustrious 18" by blackjack experts—are listed in Table 5.13 in decreasing order of their contribution to the player's edge.

The Illustrious 18 is just the beginning of the full measure of strategy decisions that call for changes to basic strategy when the count has reached a certain level. A truly proficient card counter will have memorized dozens of index numbers similar to those in Table 5.13 and be able to make the strtagey discions called for in the play of a hand. For example, in a 4-deck game where the dealer hits soft 17, the player should hit a hard 17 against a dealer 2 if the count is −20 or less [151]. With such a strong excess of low cards over high cards, the chance of busting on a 1-card draw to 17 is lower, and the chance that the dealer's hand will bust also drops considerably, so standing on 17 is a losing move. (The best strategy when facing such a strongly negative count is not to bet at all. This is a good time for a restroom break.)

When playing a game where surrender is allowed, judicious use of this option in conjunction with the true count can further add to a player's edge. In addition to the recommended surrenders in Table 5.3, the "Fab 4" surrender

TABLE 5.13: Blackjack's Illustrious 18 [105]

Strategy deviation	If the high-low true count is
Take insurance	$\geqslant +3$
Stand on 16 vs. 10	$\geqslant 0$
Stand on 15 vs. 10	$\geqslant +4$
Split 10s vs. 5	$\geqslant +5$
Split 10s vs. 6	$\geqslant +4$
Double 10 vs. 10	$\geqslant +4$
Stand on 12 vs. 3	$\geqslant +2$
Stand on 12 vs. 2	$\geqslant +3$
Double 11 vs. ace	$\geqslant +1$
Double 9 vs. 2	$\geqslant +1$
Double 10 vs. ace	$\geqslant +4$
Double 9 vs. 7	$\geqslant +3$
Stand on 16 vs. 9	$\geqslant +5$
Hit 13 vs. 2	$\leqslant -1$
Hit 12 vs. 4	$\leqslant 0$
Hit 12 vs. 5	$\leqslant -2$
Hit 12 vs. 6	$\leqslant -1$
Hit 13 vs. 3	$\leqslant -2$

decisions listed in Table 5.14, again in order of the advantage they carry, are profitable at or above the indicated count [105].

TABLE 5.14: Blackjack's Fab 4 [105]

Surrender	If the high-low true count is
14 vs. 4	$\geqslant +3$
15 vs. 10	$\geqslant 0$
15 vs. 9	$\geqslant +2$
15 vs. ace	$\geqslant +1$

It is important to understand the inherent risk here: Part of the appeal to casinos of offering the surrender option is that many players will not use it even when it's the "correct" decision. A blackjack player who surrenders correctly may be a card counter, and permitting this game choice may be part of a casino's strategy for identifying and barring counters.

The K-O Count

After the introduction of the high-low count, a number of blackjack experts set out to improve on it, whether through more sophisticated counting schemes that purported to provide more complete information or simpler schemes that were said to be easier to implement. The K-O count, named for its inventors Ken Fuchs and Olaf Vancura, is described as "the easiest card-counting system ever devised" [140]. An advantage of the K-O system is that it immediately produces the true count as play elapses, without the need to estimate the number of decks remaining or to divide by that estimate.

K-O is an *unbalanced* counting system, meaning that the total count of a complete deck or shoe is not 0. The point system for individual cards in the K-O system is shown in Table 5.15.

TABLE 5.15: K-O Count point values

Cards	Value
Low: 2, 3, 4, 5, 6, **7**	+1
Neutral: 8, 9	0
High: T, J, Q, K, A	−1

This differs from the high-low count only in that 7s are classified as low cards, with a value of +1, rather than as neutrals. With this modification, the value of a full deck is +4, and the value of an n-deck shoe is $+4n$.

The necessary twist in the K-O count is that the count at the start of a shoe starts at $4 - 4n$, where n is the number of decks in use. If you're playing a single-deck game, this means that you start at 0, as with the high-low count. In a six-deck shoe game, the count starts at −20. From this starting number, cards are counted as they appear, as in any other counting system, and the running count is updated.

Example 5.4.3. In a six-deck game with five players, the following hands are dealt:

$$A\spadesuit \ T\clubsuit, \ 7\heartsuit \ 5\spadesuit, \ J\clubsuit \ 5\diamondsuit, \ 9\heartsuit \ 7\spadesuit, \ 6\heartsuit \ 5\diamondsuit.$$

The dealer's upcard is an $A\spadesuit$. The K-O count, working from left to right, is $(-20) - 1 - 1 + 1 + 1 - 1 + 1 + 0 + 1 + 1 + 1 - 1 = -18$. ∎

The high-low count for this hand would be 0, since the count would start at 0 and the two 7s would be counted at 0 instead of +1.

Making this counting system easy is this: The true count for the K-O count is the same as the running count. No division is necessary, and there is no need to estimate the number of decks remaining. Decisions relative to changing deck composition are made by referring to the *key count* for the number of decks in play. This is a fixed number that, unlike the running count, does not fluctuate as cards are dealt. The key count is the number at which a K-O counter has an advantage over the casino [140]:

Decks	Key count
1	+2
2	+1
4	−1
6	−4
8	−6

As a comparison between the K-O count and the high-low count, consider a six-deck game. If the K-O count, which starts at −20, has risen to −4 after, say, two decks have been dealt, this means that 16 more low cards than high cards have been dealt. Two decks contain eight 7s, counted as low cards in K-O, so the average corresponding high-low running count would be +8. Dividing by the four decks remaining gives a high-low true count of +2, and this indicates a .5% player edge using basic strategy alone.

Betting in light of the K-O count is simple: For K-O novices, Vancura and Fuchs recommend a two-level wagering structure: make a "small" bet when the count is less than the key count and a "large" bet when the count exceeds this number. If the large bet is five times the small bet, a K-O counter can achieve an edge over the house ranging from .16% in an eight-deck game to .88% at a single-deck table [140, p. 77]. In addition to being simple, this betting strategy has the advantage of evading casino scrutiny, since the bet is not moving with the count, as is often done with other counting systems that call for larger and larger bets as the count grows more favorable.

For a more advanced version of K-O, there are also changes in basic strategy associated with this count. In this "K-O Preferred" strategy, insurance becomes a good bet whenever the count—with any number of decks—is at least +3. Using this modification alone adds from .16% in an eight-deck game to 1.06% at single-deck to the edges quoted above for a 1- to 5-unit bet spread [140].

+3 is, of course, the same value given for the high-low count as the benchmark for insurance bets. Consider a two-deck game in which approximately 26 cards, or half a deck, have been dealt out. If the high-low true count is +3, this means that the running count is approximately +5, indicating an excess of five low cards over high cards dealt to the table. If 26 cards have been dealt, we would expect that two 7s have appeared on average, and so the K-O count would be -4 (starting value) + 5 (for the low cards from 2 to 6) + 2 (counting 7s as low cards), or +3 again.

Under either counting system, an insurance bet would be called for if the dealer's upcard is an ace.

Since the K-O count does not divide by the number of remaining decks, additional modifications to basic strategy that are called for by the K-O count are based on index numbers. The size of the shoe is accommodated by the fact that, in some cases, these indices depend on how many decks are in play. The recommended modifications are listed in Table 5.16.

TABLE 5.16: Modifications to basic strategy based on the K-O count [140]

Decks					
1	2	4	6	8	**Strategy deviation**
+3	+3	+3	+3	+3	Take insurance
+4	+4	+4	+4	+4	Stand on 16 vs. 9
					Stand on 15 vs. 10
					Stand on 12 vs. 2
					Stand on 12 vs. 3
					Double 11 vs. ace
					Double 10 vs. 10
					Double 10 vs. ace
					Double 9 vs. 2
					Double 9 vs. 7
					Double 8 vs. 5
					Double 8 vs. 6
+2	+1	−1	−4	−6	Stand on 16 vs. 10
0	−4	n/a	n/a	n/a	Hit 13 vs. 2
					Hit 13 vs. 3
					Hit 12 vs. 4
					Hit 12 vs. 5
					Hit 12 vs. 6

These changes in playing strategy should be used as follows:

- For the "stand" and "double" decisions, deviate from basic strategy whenever the K-O count equals or exceeds the index number in the table. These decisions are correct if the shoe holds a surplus of high cards, as indicated by the count.

- For the "hit" decisions, deviate from basic strategy when the K-O count is *less than* the index number in the table. These decisions are riskier against a shoe containing a surplus of high cards, and so should only be used when the remainder of the shoe is relatively rich in low cards.

Note that the strategy associated with the K-O count includes *no* changes to basic strategy for splitting pairs or handling soft hands. Additionally, certain strategy changes are not recommended in a game with more than two decks; these are identified by "n/a" in the table.

The Big Picture

While it is not necessary to keep track of every card in the shoe to count cards successfully, that doesn't mean that that task is impossible or unhelpful—at least in theory. Under the infinite deck approximation, a more precise estimate of the effect on the basic strategy player's advantage of removing individual cards from the deck is possible [42, ?]. The player expectation E (in %) as a function of the proportions $p_1, p_2, \ldots p_{10}$ of card ranks ace through ten is

$$E(p_1, p_2, \ldots p_{10}) \approx -.69 - 52 \sum_{i=1}^{10} p_i E_i,$$

where $\boldsymbol{E} = \langle E_i \rangle$ is the 10-dimensional vector

$$\boldsymbol{E} = \langle -.59, .37, .43, .55, .69, .44, .26, .00, -.19, -.49 \rangle$$

that gives the effect of removal of a single card of rank i from a deck.

Examining this formula closely leads to the following observations:

- $p_1 + p_2 + \cdots p_{10} = 1$. These proportions are not independent: once any 9 have been specified, the tenth is automatically known. As proportions, these numbers all fall within the range $0 \leqslant p_i \leqslant 1$.

- The constant term $-.69$ arises when the expectation for a "full" deck is computed, with $p_{10} = \dfrac{4}{13}$ and $p_i = \dfrac{1}{13}$ for all other values of i. This is the HA for a blackjack game using standard rules.

- Removal of one or more 8s has no effect on the player's expectation, since $E_8 = .00$.

  Evaluation of E with $p_8 = 0, p_{10} = \dfrac{4}{12}$, and $p_i = \dfrac{1}{12}$ for $i \neq 8$ or 10, indicating a deck with all the 8s removed and all other cards present, returns the same $-.69\%$ that a full deck yields.

- As aces, 9s, or tens are removed, the corresponding proportion decreases, which decreases the negative product $p_i E_i$ and so decreases the player's expectation.

- Removal of a 2–7 raises the player's expectation by decreasing the product $52 p_i E_i$ that is subtracted from the base expectation of $-.69\%$, with the biggest increase coming from removing a 5—as Thorp's five-count initially surmised.

This information could be used to derive the ultimate card-counting system, which would assign appropriate values to every rank in the deck that measure their precise significance to players. This system uses the point values shown in Table 5.17. As might be expected, this system achieves great accuracy only at the cost of increased complexity.

A careful observer cannot fail to see how the high/low counting system is a fair approximation to these numbers. With the exception of the values for 5 and 7, the high/low system uses these numbers with standard arithmetic rounding rules applied. (Another unbalanced counting system, the *Red Seven* count, counts red 7s as +1 and black 7s as 0, making ½ the average value of a 7 and coming even closer than high/low to these ultimate values.)

TABLE 5.17: Ultimate blackjack counting system [145]

Card	A	2	3	4	5	6	7	8	9	T
Value	−1.28	0.82	0.94	1.21	1.51	0.98	0.57	−0.06	−0.42	−1.07

A card counter who is adept at adding and subtracting numbers with 2 decimal places under live casino conditions can take this count system into a casino and be assured that he or she is playing the most accurate game of blackjack possible. Deriving the strategy that corresponds with this system remains an unexplored challenge.

Casino Pushback

It must be stressed at the outset here that *card counting is not illegal.* Card counters are simply using their brains to keep track of what's going on at the table, and it is difficult to imagine the magnitude of the fallout from a law forbidding thinking. However, while they are violating no laws, that does not give them free rein to practice their craft. Casinos, being private businesses, are free to restrict or forbid access to anyone they choose, and so casinos in Nevada may, if they wish, eject suspected card counters.

New Jersey law, by contrast, forbids casinos from barring card counters; Atlantic City casinos deal with counters by restricting the amount they are allowed to bet or by shuffling after every hand, and so eliminating any advantage that might be gained from counting cards. Bob Stupak followed a similar policy at Vegas World, where he was known to issue identified card counters a card stating that they were welcome to play blackjack, but that their wagers were to be restricted to no more than 7 units [127]. This meant that, for example, at a $5 table, known counters could not bet more than $35.

Casino reactions to card counting seek an elusive balance between the fact that successful card counters have the potential to win a lot more money than the house advantage says that they should and the equally important fact that far more people think they can count cards successfully than actually can, which leads to increased traffic at blackjack tables and, ultimately, the likelihood of increased casino income from the tables. Since the first of these facts is easier for casinos to address, we begin by describing some casino countermeasures.

Rule Changes

One of the first things that casino managers did after *Beat the Dealer* was published was to tighten up the rules of the game in an effort to eliminate the advantages of card counting. On April 1, 1964, the Las Vegas Resort Hotel Association announced that henceforth, double downs would be allowed only on hands of 11 and that blackjack players were forbidden to split aces [135].

These changes eliminated two very advantageous situations for card counters, since a high true count favors doubling down on a 9, 10, or certain soft hands against an unfavorable dealer hand, and splitting aces, which is always the right basic strategy move, is even more lucrative when the remaining deck is rich in 10s. Table 5.1 can be used to calculate the effect of these changes.

- Double down only on 11: +0.78%.
- No splitting aces: +0.18%.

The HA with these new rules went up by .96%, but perhaps the bigger effect was on the public's perception that casino operators were explicitly disadvantaging talented players. Blackjack players—counters and noncounters alike—stayed away from the games with new unfavorable rules, and they were reversed within three weeks.

After that reversal, casinos started looking for other game modifications that might restore their advantage and started moving from single-deck hand-dealt games to multiple-deck shoe games. Adding decks, in addition to increasing casino profits because more time is spent dealing the cards than shuffling them, increases the HA, but this increase levels off as more decks are added.

- Two decks: +0.32%
- Four decks: +0.48%
- Six decks: +0.54%
- Eight decks: +0.57%

The casino derives no significant game-related advantage in going past eight decks, and the greater height of the stack of cards may make it unwieldy for a dealer to handle.

Card counters then responded by developing new guidelines for counting into multideck shoes, which led to the running count and true count, and so the arms race of sorts between players and casinos was on. As casinos found a new rule or game procedure that would improve their position, the card-counting community developed a modification to their tactics that would combat it, and vice versa.

Since one important part of card counting is bet fluctuation—you bet more when the count is highly positive to take advantage of favorable conditions—casinos began barring players whose bets went up and down too widely over a period of observation, or using *preferential shuffling*: shuffling the cards as soon as a player increased their bet. This worked both ways, of course—a counter could increase his or her bet during a highly *negative* deck to encourage the dealer to shuffle early and give up the house's edge. At many casinos, dealers are not part of the defense against card counters, as they have enough to do just running the game properly. As a result, the dealer might well have been unaware of the count and simply reacted to the jump in the player's wager.

Even nonpreferential shuffling sometimes works in a casino's favor. In a single-deck game with five players against the dealer, one can safely expect two full rounds to be dealt, since an average blackjack hand uses 2.7 cards. With six hands per round, we would expect 33 cards to be dealt in the first two rounds, leaving only 19 for a third round. Many dealers will shuffle the cards when less than half a deck remains, so two rounds is about all one would expect—unless those first two rounds use an unusually small number of cards. When would that happen? Precisely when most players stand on their initial two-card totals, and this frequently happens when the player's hands are rich in high cards, leaving a partial deck with a highly negative count, favoring the casino, on the third round. This benefit accrues to the casino whether or not the players are counting cards.

Electronic blackjack games which use from 1 to 8 decks commonly reshuffle the cards after every hand, making card counting at these machines a pointless exercise, even thought the lack of close supervision might make counting seem like a profitable pursuit.

A significant player reaction to casino action against bet variation was the idea of team play, where certain team members would play and count while flat-betting, then when the count was high, call in a "big player" who came in, made a few large bets, and left when the count dropped. Since the big player simply made large bets at advantageous tables, there was no bet variation to arouse suspicion—to all appearances, he or she was just a lucky high roller moving among tables. Some of the most famous blackjack teams were based at the Massachusetts Institute of Technology (MIT); their exploits were chronicled by Ben Mezrich [79, 80] and portrayed in the movie *21*.

Casinos countered team play by forbidding mid-shoe entry, requiring players to wait for the shuffle before joining a game and thus blocking "big players" from jumping into favorable shoes. This also protected the casino against *wonging*: the practice where an individual player counts down a deck or shoe while standing behind the table instead of wagering, then settles in to play only if the count grows favorable.

At the same time, the publication of the Illustrious 18 and other recommended deviations from basic strategy led casinos to watch players somewhat more closely for such deviations. A player who splits tens when the deck is highly positive (part of the Illustrious 18 if the dealer's upcard is a 5 or 6), for example, is almost certain to attract scrutiny from casino personnel, for this is a move that is almost never made by noncounters. A player who splits 10s is either very inept—and thus a valued customer—or a card counter. Ongoing observation of the player's actions relative to the count—and players should take it for granted that *someone* on the casino security staff knows how to count cards—will easily determine which. The casino can then act accordingly: encouraging the inept player with complimentary meals, hotel rooms, or other inducements, and barring the counter or initiating countermeasures to make the game less attractive.

Rise of the Machines

In the early 1970s, some card counters turned to the new technology of home-built portable computers to automate and expedite the counting process. Technical restrictions hampered many of these devices, which had to be small enough to be undetectably concealed on a player's person. Some were controlled through switches operated by a player's toes, and these were frequently unreliable. Additionally, some means had to be found to signal the correct playing decision to the player, and this posed another significant challenge. Radio transmitters ran the risk of detection when they interfered with casino security systems, and smaller devices that connected to fake digital watches or LEDs imbedded in eyeglass frames were tricky to conceal. This age of innovation emphatically ended in Nevada on July 1, 1985, when portable computers were banned in casinos. Violators of this law are subject to fines of up to $10,000 plus 1 to 10 years in jail [139]. New Jersey followed with its own law, and as new gaming jurisdictions came online, they swiftly moved to ban card-counting computers.

This issue resurfaced beginning in 2009, when the introduction of Apple's iPhone led several developers to program and market card-counting applications. Some of these apps were designed to operate even when the screen was turned off, and so avoid the giveaway of an illuminated screen. Casino operators were warned about the apps, and the Nevada Gaming Control Board swiftly forbade their use in a casino. Card counting may be legal, though not exactly welcome, but using an electronic device to count cards is emphatically not.

One more recent device introduced by casinos to counteract card counting is the *continuous shuffling machine* (CSM). A CSM is a machine that collects the discards from each hand and immediately re-inserts them into the packet of cards, thus effectively dealing every hand from a fresh shoe of four to eight decks and eliminating all of the advantages of card counting. As when casinos addressed the advent of card counting by changing the rules to restrict player actions in the 1960s, CSMs as a casino response to advantage play proved to be something of an overreach. Part of the continuing appeal of blackjack is the fact—and it is a fact—that it can be beaten with sufficient work. Continuous shuffling machines, by completely negating the advantages to be gained from counting cards, demolish this selling point while adding .30% to the HA. While CSMs are still to be found on some casino floors, serious gamblers avoided them so assiduously that blackjack income dropped off. In a competitive market such as Las Vegas or Reno, gamblers had options, and exercised them. In a market with a small number of casinos such as Osceola, Iowa—home to only one casino, the Lakeside—blackjack players must take the game as offered and so have little recourse to resist any anti-counting strategies. Of course, smaller markets such as Osceola are less susceptible to major attacks by card counters.

Card counters are obviously disadvantaged by CSMs, but do they affect ordinary basic strategy players? The play of an individual hand is not affected for noncounters; the effect kicks in to their detriment when we consider the increased game pace that CSMs allow. In light of the fact that no playing time is lost to shuffling the cards, a CSM allows the dealer to deal more hands per hour than an ordinary shoe, and this exposes the players to more hands where the casino's advantage works against them.

It should be noted that there's a difference between a CSM and an automatic shuffler. Many casinos have turned to shuffling machines in order to reduce the time spent not dealing blackjack, but only if the cards are continuously replaced in the shoe are you dealing with a CSM. If the used cards are collected in a tray for multiple hands before being inserted into the machine, and if there's a pause in the game while the cards are machine shuffled, then the only disadvantage to players—counters or not—comes from the aforementioned increased number of hands dealt per hour.

The Aladdin Strikes A Balance

With regard to the second fact described above, the tendency of many casino patrons to overestimate their ability to count cards, casino management simply needs to find a way to welcome these players, who pose no real threat to their blackjack tables, for low-level gambling that won't cost the casino too much in the event of a short-term run of player luck. This process includes the realistic assessment of anti-counter methods for their effect on the casino's overall financial health. Casino management expert Bill Zender has advised casinos against overreaction to the supposed threat posed by counters, claiming that excessive countermeasures had the unintended effect of decreasing the casino's profits by wasting time trying to respond to an overstated threat. Zender's assertion is that increasing the number of blackjack hands dealt per hour is key to improving a casino's bottom line, and such tactics as excessively complicated shuffles (to deter "shuffle trackers," an advanced type of player who tries to track the movement of groups of cards through the shuffle and bet appropriately when the same cards come up in the next round [80]), barring mid-shoe entry, and dealing less than 80% into a shoe (75% in hand-held single- or double-deck games) work against maximum dealer efficiency [153].

As casino manager at the Aladdin Casino in Las Vegas (now Planet Hollywood) during the mid-1990s, Zender instituted a number of policies that, taken together, had the effect of producing one of the best blackjack games offered on the Strip at the time [78]. The Aladdin offered the following rules on its blackjack tables [153, p. 17–22]:

- *Single-deck games replaced by double-deck games.* This adds .32% to the casino advantage. Zender's contention was that players don't readily distinguish between the two games, classifying them all as "hand-held." Since single-deck blackjack is approximately an even game for players using perfect basic strategy, and is specifically sought out by expert players, this move to two decks benefits the casino without raising too many player objections. The objections that were raised at the Aladdin came from the more knowledgable players, the loss of whose business was not necessarily regarded as tragic.

- *Hitting soft 17.* We have previously looked at the reasons why a casino would do this, which has become standard practice across the Las Vegas market. Zender noted that many casino patrons actually favored this change that gives the casino an additional .2% edge, since it gives them a second chance to beat the dealer when they stand on a low hand.

Example 5.4.4. If the dealer's upcard is a 6, basic strategy calls for a player to stand on 12 or higher. If the player stands on a 13, and the dealer then turns over an ace for a soft 17, a losing hand for the player is turned into a hand with another chance to win. ∎

- *Dealing 5½ decks from a six-deck shoe, and 1½ decks in a double-deck game.* This was a change that was difficult to sell to floor personnel, due to their overestimate of the threat posed by card counters. Nonetheless, Zender sold it as a way to spend more time dealing and less time shuffling, and the added casino revenue per six-deck table per year due to this switch is in the hundreds of thousands of dollars.

- *Offering late surrender.* While this option gives players a .06% edge over the casino, the fact that few players exercise it, and fewer still use it correctly (Example 5.2.2), makes this a relatively inexpensive rule to implement, and it can be explained as compensation to players for hitting soft 17s. What the casino receives from this rule change is insight that might help identify card counters, since most players who surrender correctly are doing so in conjunction with knowledge of the count.

- *Ignoring card counters whose top wager was $50 or less.* Such counters, whatever their skill level, pose very little threat to the casino's bottom line. This change established the Aladdin as a counter-friendly casino, and numerous aspiring counters—of all skill levels—descended on their games, raising the casino's income. Since the number of truly skilled card counters is wildly overestimated, and the number of novice counters prone to errors is large, this change benefited both parties.

Pessimistic observers confidently predicted that card counters would exploit the new procedures to their fullest advantage and to the detriment of the casino [78]. The reality was far different: The net effect of these changes was that the Aladdin's blackjack revenue during this period was consistently 2% over the average for the Las Vegas Strip [153].

5.5   Additional Topics in Card Counting

Counting Over/Under

Blackjack side bets often have a high house advantage to offset the low edge the casino holds against a basic strategy player. Over/Under proved to be an exception.

The infinite deck approximation used to analyze that bet (Table 5.4) conceals, perhaps, the fact that Over/Under is susceptible to card counting. An early study using the High-Low count in a 6-deck game showed that the Over bet had a player advantage at a true count of +5 and the Under bet shifted to favor gamblers when the true count reached −8 [128]. This was a case of adapting an existing count system to a different wager, and while it was adequate for identifying a player edge, a specialized counting system tailored to Over/Under was soon developed as the game spread more widely from northern Nevada.

Of particular significance in the Over/Under count was proper treatment of aces, which count as low cards for Over/Under but are counted with the 10s in High-Low. The Over/Under count uses the following values [129]:

Cards	Value
Low: A, 2, 3, 4	+1
Neutral: 5, 6, 7, 8, 9	0
High: 10, J, Q, K	−1

With the Over/Under count, players have the advantage on the Over bet when the true count is +3 or greater, and have the edge on Under at −4 or less [129].

Card Counting in Baccarat

Baccarat, of course, is also a game in which cards are not replaced after each hand, and so prior hands affect future hands. This naturally leads to the question of whether card counting as employed in blackjack can be used in baccarat to improve a player's position relative to the casino. The answer, perhaps surprisingly, is "no." There are a number of reasons for this:

- Baccarat play is rigidly prescribed, with no place for player decisions. Card counting often calls for different player action under different deck compositions, including such actions as doubling down or splitting pairs that are not available in baccarat.

- An important part of the strategy behind card counting in blackjack involves standing on a low hand when the deck is rich in high cards in the hopes that the dealer will bust. Since baccarat hands do not bust if they go over 9, there is nothing to be gained here by tracking specific cards. Ten-count cards in blackjack count zero in baccarat; thus there's no real advantage in tracking how many remain in the shoe.

- There is no bonus for a natural in baccarat, whereas a natural 21 in blackjack pays better than even money (3 to 2 or 6 to 5, depending on the casino). Knowing when natural 8s or 9s are more likely to occur might have some value as far as increasing a wager, but even a natural pays only 1 to 1.

- The bettor may wager on either the Banker or the Player hand. Since the rules are approximately symmetric for the play of the two hands, there are no cards that can be said to be more favorable to one side than the other, as is the case with tens in blackjack, an excess of which favors the player over the dealer.

Not long after the introduction of card counting methods to blackjack, mathematicians naturally examined how information gained from counting cards might be used to gain an edge at baccarat. Edward Thorp soon concluded that "no practical card counting systems are possible" for baccarat [42]. Given a particular block of six cards ready to be dealt, it is possible for the sharp player to gain an advantage. For example, if it is known that the next six cards to be dealt are all 7s, which leads to both hands totaling 4, drawing a third card, and tying at 1, a large Tie bet is guaranteed to win—but it is extremely unlikely that any baccarat player would encounter such a setting. Is it possible to count cards at baccarat? Of course—many card counting systems have been developed. Does it matter? No—the advantage that you might gain is, at best, microscopic. One source claimed a 15¢ advantage per shoe for a player betting $1000 when the shoe turned favorable, an edge of .015%, which is hardly worth the effort or the risk of exposure [56].

Nonetheless, most casinos are happy to provide scorecards to players who wish to track the winning hand from deal to deal in an effort to find and exploit patterns. The patterns certainly exist, but this is only because in an extended gaming session, there will always be apparent patterns in the sequence of winning hands. For all practical purposes, the winning hand in a round of baccarat is independent of the previous rounds' winners.

However, side bets might be a different matter. Some baccarat tables once featured the *Banker 9* side bet, a separate wager that paid off at 9 to 1 if the Banker hand was a natural 9. Thorp and William Walden reported that the probability of winning this wager was .0949 in an eight-deck game [136]. The corresponding expectation is

$$E = (9) \cdot (.0949) + (-1) \cdot (1 - .0949) = -.0510,$$

which gives the casino a 5.10% edge.

The dependence of any one hand on the cards previously dealt means that the probability p of winning this bet is not fixed. A look at the expectation for the Banker 9 wager shows that the bettor has an edge if $10p - 1 > 0$, or if $p > .10$. This suggests that careful tracking of the cards remaining in the shoe might reveal a point where the bet favors the gambler. In light of the fact that there are many ways to draw a two-card total of 9, this might seem like too complicated an undertaking, though, so a simple approach focuses on the most common combination totaling 9: a dealt 9 with a ten-count card.

Thorp and Walden then proceeded to outline a strategy that would lead to a player advantage for the Banker 9 wager. The strategy involves counting n, the total number of cards, and t, the total number of 9s, still in the deck. In a fresh eight-deck shoe, $n = 416$ and $\frac{n}{t} = 13/1$; as $\frac{n}{t}$ decreases, indicating an excess of 9s over the other cards in the shoe, the Banker 9 bet becomes more advantageous for the gambler. The ratio $\frac{n}{t}$, combined with the value of n, determines the optimal amount to wager, as a fraction of a gambler's total bankroll V, on the Banker 9 bet. For example, if $n \approx 130$ and $\frac{n}{t} = 5$, the optimal strategy calls for a wager of 4.7% of V [136].

This was terrific in theory, but was it practical for casino use? A team of players tried the system in two Nevada casinos, winning $100 per hour for seven nights in one and $1000 per hour for two hours in the other before being barred from both. The net result of this discovery was that Nevada casinos discontinued the Banker 9 side bet [135].

5.6   Exercises

Answers to starred exercises begin on page 289.

5.1.* Consider the following scenario: On the first round of a single-deck blackjack game, there are three players. You are dealt $K\diamondsuit Q\clubsuit$, player #2 is dealt $2\clubsuit 7\spadesuit$, and player #3 is dealt $6\heartsuit 9\spadesuit$. The dealer's upcard is the $3\clubsuit$. You naturally stand on your 20. Player #2 doubles down on her 9 and draws the $J\clubsuit$ for a 19. Player #3 stands on his 15, as basic strategy directs. The dealer turns over the $4\clubsuit$ and draws the $K\spadesuit$ for a 17, and so the first hand ends with you and player #2 winning and Player #3 losing.

On the next hand, you have bet $10 and the dealer's upcard is the $A\diamondsuit$. You have $A\spadesuit T\diamondsuit$, a natural 21. Based on this information, is a $5 insurance bet a good idea? Explain your answer by computing the expected value of the total bet including insurance, bearing in mind that if the insurance bet wins, your original bet pushes, and if the insurance bet loses, your original bet pays off at 3 to 2.

5.2. Some casinos have offered an insurance bet when the dealer's upcard is a ten-count card, which pays off at 10 to 1 odds if the hole card turns out to be an ace.

a. Assume that the game is being dealt from a fresh double deck and that the only cards you can see are your hand and the upcard. Calculate the expected value of this bet if you hold

   i. No aces.

   ii. One ace.

   iii. Two aces.

b. If you have a natural, calculate the expectation of your total wager, assuming that you make an insurance bet for half of your original bet and that blackjack pays off at 3 to 2. How does it compare to the expectation without an insurance bet?

5.3.* Use the infinite deck approximation to estimate the probability of winning the Lucky Lucky side bet (page 225).

5.4. The Four Queens Casino in downtown Las Vegas once offered a "Red/Black" blackjack side bet on the color of the dealer's upcard [148]. The bet paid off at even money, with the provision that if the upcard was a deuce of the color bet, the wager pushed. Find the expectation of this bet.

5.5.* Basic strategy for Double Exposure 21 calls for a player to hit a hard 20 if the dealer is also showing a total of 20 and hope for an ace, since ties lose. Use the infinite deck approximation to find the probability that both the player and dealer are dealt hard 20s.

5.6. Single-deck blackjack under reasonable rules, including a 3 to 2 payment on naturals, is nearly an even game, and for that reason, casinos tend to scrutinize play at single-deck tables more closely. A technique called *depth charging* has been used by some high rollers to try and gain an advantage at single-deck games. The gambler plays all seven hands on the table, betting increasing amounts on each hand. For example, the wager on the first hand may be $100, then $200, $300, and so on, up to $700 on the last hand to receive cards. In practice, single-deck blackjack is dealt with all player cards face down, and players playing multiple hands may not look at any hand until all previous hands are completely played out.

How does a player gain any advantage from depth charging?

5.7. Blackjack Switch offers a side bet called the *Super Match* bet, which pays off if the player's first four cards contain a pair or better. The payoff table for the Super Match bet is

Combination	Payoff
One pair	1 to 1
Three of a kind	5 to 1
Two pairs	8 to 1
Four of a kind	40 to 1

Assuming a hand that's being dealt from the top of a fresh six-deck shoe, show that the house advantage on this bet is approximately 2.55%.

5.8. The *Perfect Pairs* side bet, seen at the MGM Grand Casino in Detroit, pays off the following amounts if the player is dealt a pair in his or her first two cards:

- **Mixed Pair**: If the pair contains cards of different colors, the payoff is 5 to 1.

- **Colored Pair**: If the pair consists of two cards of the same color but different suits, the bet pays 10 to 1.

- **Perfect Pair**: A pair consisting of two identical cards (for example, two 7♢s) pays 30 to 1.

For a blackjack game dealt from a five-deck shoe, calculate the house advantage for Perfect Pairs.

5.9.* The Tropicana Casino in Las Vegas once offered a blackjack bonus for a hand consisting of four red 5s [148]. For a game dealt from a six-deck shoe, calculate the probability of being dealt such a hand. How does this probability change in the infinite deck approximation?

5.10.* For a Jack-A-Tack bet, use the infinite deck approximation to estimate the probability of winning the top 299–1 prize..

5.11. In a four-deck game without surrender, 100 cards have been dealt out, in accordance with the following table:

$$
\begin{array}{ll}
\text{Low cards:} & 45 \\
\text{7s:} & 4 \\
\text{8s or 9s:} & 13 \\
\text{High cards:} & 38
\end{array}
$$

On the next hand, you have been dealt 9♡7♠. The dealer shows the T♢.

a. Compute the high-low count and determine the correct play for this hand.

b. Compute the K-O count and determine the correct play for this hand.

5.12. The Instant 18 side bet (page 214) gives a player making it a hypothetical hand with a value of 18. In [54], we find the following distribution for the final value of the dealer's hand in a single-deck game where the dealer stands on soft 17:

x	17	18	19	20	21	> 21
$P(x)$	.1458	.1381	.1348	.1758	.1219	.2836

Using this distribution, find the expectation of a \$1 Instant 18 wager.

5.13. Consider a single-deck blackjack game with a full table of seven players, plus the dealer. Given that the average blackjack hand uses up 2.7 cards, find the probability that all four aces will be dealt on the first round, leaving no possibility for a natural in the second round.

5.14. If you're playing single-deck blackjack and are dealt a pair of 7s against a dealer 10, a variation on basic strategy specific to one-deck games states that you should stand on your 14 rather than hitting, as called for in Table 5.6. In light of the fact that the dealer's hole card is more likely to be a 10 than any other value, explain this strategy change.

5.15. A possible wager that could be added to Over/Under is a "13" bet, that the player's hand will total exactly 13.

a. What is the HA if this bet pays 10–1?

b. If the 13 bet pays off at x to 1, find x so that this bet is fair.

5.16. *21 Superbucks* is a blackjack side bet that appeared at the Treasure Bay Casino in Biloxi, Mississippi [49]. The side bet pays off if the player's first two cards are suited and in sequence. Find the probability of winning this bet if the game is dealt from a 6-deck shoe.

5.17. The Royal City Star Casino in New Westminster, British Columbia offered a $1 blackjack side bet that paid off on the number of aces in a player's hand. A progressive jackpot was paid if a player was dealt 4 aces of the same color [44]. In a 6-deck shoe, find the probability of receiving 4 aces of the same color.

5.18. The *Lucky Aces* blackjack side bet pays off based on the first 4 cards dealt to the player and the dealer. Table 5.18 shows the pay table.

TABLE 5.18: Lucky Aces pay table

Cards	Payoff
4 aces	500–1
3 suited aces	250–1
3 aces	20–1
2 suited aces	8–1
2 aces	5–1
1 $A\diamondsuit$	2–1
1 ace	1–1

Since this bet is offered on an electronic blackjack machine using a 6-deck shoe that is reshuffled after every hand, computing winning probabilities can be done without regard for previously-dealt cards, as we saw with the Bonus blackjack side bet. Find the probability of each event listed in Table 5.18.

Chapter 6

Betting Strategies: Why They Don't Work

Here's an inescapable mathematical fact:

> No betting strategy can overcome a negative expectation.

This is because expected value is additive; that is,

$$E(X_1 + X_2) = E(X_1) + E(X_2),$$

as established in Theorem 3.2.1. Translated into English, this means that if the expectation of a bet is negative, making bigger bets, or multiple bets together, will still produce a negative expectation. A more mathematical way of assessing multiple bets is this: The expectation of a collection of simultaneous bets can never be better than the expectation of the most favorable bet or worse than the expectation of the least favorable bet.

This does not stop people from developing and marketing "get rich quick" schemes for pretty much any casino game. In this chapter, we will look at some of the more common ruses either intentionally or unintentionally offered for sale to unsuspecting gamblers.

6.1 Roulette Strategies

There are numerous strategies proposed for roulette, perhaps more than for any other casino game. This is likely because roulette is easy to understand and to play, and thus systems for roulette carry the lure of easy money with little effort. By contrast, blackjack gives an impression of requiring considerable effort to master, and the craps layout is complicated and can intimidate new gamblers.

Example 6.1.1. Here's a simple strategy for roulette which was first published in a Cuban magazine, *Bohemia*, in 1959 and further described by gambling scholar John Scarne [104].

DOI: 10.1201/9781003358183-6

> *Bet one chip on black and one on the third column, which, owing to a fluke of the layout, is unbalanced and contains only four black numbers among its dozen.*

The usual claim here revolves around the idea that this bet covers 26 of the 38 slots on the wheel, and 4 of them are covered by both bets. Considered only at that level, this may sound promising, but we need to consider the expected value of the two-unit bet before passing our final judgment:

- If one of the four black numbers in the third column (6, 15, 24, and 33) comes up, both bets win, and we profit $3.

- If one of the eight red numbers in the third column comes up, we win the column bet at 2 to 1 but lose the color bet at 1 to 1, for a net win of $1.

- If one of the 14 black numbers in the first or second column comes up, we break even, losing the column bet but winning $1 on the color bet.

- This leaves only 12 numbers—10 red and 2 green—on which we lose $2.

All possibilities are covered here, and so the expected return is

$$E = (3) \cdot \frac{4}{38} + (1) \cdot \frac{8}{38} + (0) \cdot \frac{14}{38} + (-2) \cdot \frac{12}{38} = -\frac{2}{19},$$

and the standard HA of 5.26% has returned. This collection of bets, each of which has an HA of 5.26%, combines to produce a compound bet with the same house edge. Note here that since we bet $2, a return of $-\$\frac{2}{19}$ represents a loss of 5.26% of our original wager.   ■

Needless to say, this was not the mathematical analysis provided by the original purveyor of this scheme.

Martingale

The *martingale* or *double-up* strategy is rediscovered from time to time, and is most often employed on even-money roulette propositions such as red or black. It can also be used in craps, blackjack, baccarat, or any other game where the probability of winning a bet is close to .5.

> *Bet $1 on the first spin of the wheel. If you win, pocket your $1 profit. If you lose, bet $2 on the next spin. If you win, your net profit is $1—quit while you're ahead. If you lose again, bet $4 on the next spin—win and you walk away with $1, lose and your next bet is $8, and so on. In short, you double up after every losing spin. Since you will win eventually, when you do, you'll walk away a $1 winner.*

Leaving out the concern that this seems like a lot of effort in order to win $1, let's see what we have here. Every statement in this scheme is true until the very end. It is true that you will win "eventually" (the probability of an infinite string of losses is 0), but that can take a very long string of spins. For example, if you bet on red and red fails to come up ten times in a row, an event with probability $\left(\frac{20}{38}\right)^{10} \approx .001631 \approx \frac{1}{613}$ (small, to be sure, but not impossibly so), then your losses total $1023, and your 11th bet would be for $1024 in an attempt to wipe out all of your previous losses and come out $1 ahead. That's a lot to risk for such a small payoff—and you are still more likely to lose than win that bet. If you lose that one, you're down $2047 and are next supposed to put $2048 on the line to come out $1 ahead.

But there's another problem that arises. Casinos typically have maximum bet limits on their roulette tables. For example, suppose that your chosen roulette table displays the sign in Figure 6.1.

ROULETTE $1 GAME

INSIDE BETS	OUTSIDE BETS
$1 MINIMUM	$5 MINIMUM
$100 MAXIMUM	$2000 MAXIMUM
ANY BET	PAYOUT

PLEASE REDEEM ROULETTE CHECKS HERE

FIGURE 6.1: Roulette table sign showing bet limits.

This sign refers to two different bet limits:

- On "inside bets," which are bets made on the numbered part of the layout covering from one to six numbers, such as a single-number bet on 23 or a double street bet covering 1 to 6, the minimum bet is $1 and the maximum bet is $100. The casino's maximum payoff on any winning inside bet is limited to $3500.

- On "outside bets," which are placed along the edge of the layout on even-money and 2 to 1 bets such as red/black or the three columns, the minimum is $5 and the maximum *payout* is $2000. The limit on the payout effectively limits even-money outside bets to $2000 and 2 to 1 outside bets to $1000.

The limits are higher for outside bets because the payoff odds are lower, so the casino is exposed to less risk. For the martingale, you will be making outside bets.

If you begin with a $5 bet (the minimum for outside bets) on red, then after nine straight losses, the table limit of $2000 on payouts will prevent you from making the next bet, for $2560, that will recoup all of your losses and give you a $5 profit. Since you're making even-money bets, the maximum of $2000 on payouts is also the limit on your martingale wager.

Suppose that you begin with a $1 bet at a table with a $1000 maximum on outside bets and play this system exactly 1024 times betting on red, where your probability of winning on each spin is approximately .4737. In 485 games, you'll win your dollar on the first spin. On 255 more, you'll lose $1 on the first spin but then win back $2 on the next for a net win of $1. You'll lose on the first two spins before winning on the third in 134 games—again profiting by $1—and so on until game 1022, when you lose the first nine spins, dropping $511 before making it all back with a winning $512 bet on the tenth spin. Through the first 1022 games, you'll be ahead $1 on each, for a total of $1022. On games 1023 and 1024, though, you lose on the first ten spins, for a total loss of $1023 and no opportunity to make the 11th bet that wipes out all of your losses. The best you can do at that point is to bet $1000 and hope to win that bet, which will leave you down only $23. Lose that bet, however, and you'll find yourself down $2023 and still unable to bet more than $1000.

A more accurate description of this martingale would be "When you win, you'll win $1 each time. But when you lose—and if you use this strategy often enough, you will eventually lose—you'll lose around $1000." The system would be foolproof if you had an infinite bankroll and the casino had no bet limits—although if you had an infinite amount of money, what would be the point of gambling? The thrill of the game, by itself, would seem to be insufficient motivation.

Fibonacci Progression

The *Fibonacci progression* attempts to build a system out of the Fibonacci sequence of integers [4]. The first two numbers in the Fibonacci sequence are $F_0 = 0$ and $F_1 = 1$, and each successive term of the sequence is found by adding the two previous numbers together. Thus, the sequence continues $F_2 = 0 + 1 = 1, F_3 = 1 + 1 = 2$, and so on. The sequence begins

$$0, 1, 1, 2, 3, 5, 8, 13, 21, 34, 55, 89, 144, \ldots$$

and can be extended as long as desired.

A roulette system based on the Fibonacci sequence works like the martingale. You begin by betting $1, but then instead of doubling your bet after each loss, you bet the number of dollars given by the next number in the Fibonacci sequence. After you win, instead of returning to a $1 bet, you bet the previous number in the sequence, and so move your way up and down the list as the spins accumulate. Owing to the nature of the sequence, you bet somewhat less than twice your previous bet after you pass a $2 bet; the increase between successive bets after a loss is approximately 61.8% rather than 100%.

Nonetheless, this suffers from the same flaw as the martingale—following a stretch of losses, you will be betting a large sum in order to come out a small amount ahead—you'll just lose somewhat more slowly and hit the casino's maximum bet limit after a few more bets than with the martingale. Using the standard martingale strategy with a $1 minimum and $1000 maximum wager allows you to make ten bets before hitting the maximum; the Fibonacci strategy can endure 16 straight losses before the next bet, for $1597, would exceed $1000.

Kryptos

The *Kryptos* system for roulette is a variation on the martingale that attempts to account for the independence of successive wheel spins. The idea behind Kryptos is that there are three attributes to every nonzero roulette number: high/low, odd/even, and red/black. As described in [65],

> *Here, what we are trying to do is forecast, at random, the next spins of the wheel, counting on the fact that we really can't forecast them correctly.*

Six spins of the wheel are recorded, in an effort to secure a random sample. Suppose that those spins are 9 (low, odd, red), 11 (low, odd, black), 33 (high, odd, black), 21 (high, odd, red), 36 (high, even, red), and 6 (low, even, black). We are interested only in the attributes of the numbers, and not the numbers themselves, and so we arrange those attributes in a table:

L	O	R
L	O	B
H	O	B
H	O	R
H	E	R
L	E	B

The argument in support of Kryptos is that it's highly unlikely that all six of these patterns will be completely wrong, thus making it highly likely that betting the even-money chances as listed will result in one win per column by the end of the sixth spin. We now make three separate even-money bets for $1, on low, odd, and red—corresponding to the first row. If a bet wins, we collect our profit and cease betting that column until all six rows are complete. If a bet loses, we double the bet on the choice indicated in that column of the next row (a martingale system). For example, if we make our three bets and the first number then spun is 2 (low, even, and black), we take down $1 of winnings on low and bet $2 each on odd and black (row 2) for the next spin. If that spin yields the high/odd/red 21, we collect $2 from the odd bet and next bet $4 on red. The next spin is a black 11, so our next bet is $8 on red. That spin is a red 19, so we collect $8. Our accumulated losses come to

$1 + 1 + 2 + 4 = \$8$, while our wins total $1 + 2 + 8 = \$11$—a \$3 profit, or precisely the payoff of three separate martingales.

Clearly, the risks of Kryptos are that one or more of the three bets will lose six times in a row, or that the table limit on even-money bets will be reached. The probability of losing six straight even-money bets at American roulette is

$$p = \left(\frac{20}{38}\right)^6 \approx .0213$$

—so there's about a 2% chance that any one column will lose six times in a row and cost \$63. While the author of this system claims that "this will be more than offset by our winnings," it will take a lot of \$3 winning sessions to make up such a setback.

Cancellation

The *Cancellation* or *Labouchère* system is often suggested for roulette, but it can be used for any game with approximately even-money wagers [104].

> *If your goal is to win \$N, write down a sequence of numbers that add up to N. For your first bet, add the top and bottom numbers together and wager that amount. If you win, cross the numbers out. If you lose, add the amount you lost to the bottom of the list. For the next bet, add the top and bottom numbers and bet that amount, and continue in this manner.*
>
> *Every time you win, you cross off two numbers. Every time you lose, you add only one. Therefore, you will eventually cross off all the numbers, at which time your profit is the \$N you sought.*

An immediate problem here is that a gambler using this system will be making increasingly larger bets as the gaming session goes on. Consider the following sequence of wagers and spins, where you are out to win \$36:

Original list: {1, 2, 3, 4, 5, 6, 7, 8}
Assume that you always bet on red. The first bet is for 9 units, and loses.
Second list: {1, 2, 3, 4, 5, 6, 7, 8, 9}
The second bet is for \$10 and wins. The net profit is \$1.
Third list: {2, 3, 4, 5, 6, 7, 8}
The next bet is for \$10 and loses. The net loss is –\$9.
Fourth list: {2, 3, 4, 5, 6, 7, 8, 10}
The fourth bet is for \$12 and wins, for a net win of \$3. The list now reads {3, 4, 5, 6, 7, 8}. Note that the amount you have won is equal to the sum of the numbers you have crossed off the original list—the \$9, \$10, and \$12 bets were just means to this end.

Suppose that the next three bets—for \$11, \$14, and \$17—all lose. The list now reads {3, 4, 5, 6, 7, 8, 11, 14, 17}, you are down \$39, and your next bet is to be for \$20.

If your luck is average or worse, the net effect of the Cancellation System is that you will be betting large amounts in order to win small amounts, for you'll never be ahead more than the sum of the numbers that you originally wrote down. At some point, then, the very real risk that you will run out of money comes into play, and no system works if you can't afford the wager.

The Gambler's Fallacy Resurfaces

An apparently self-published pamphlet called *Famous Las Vegas Gambling Systems* [26] purports to offer winning systems for a range of casino games. Several of the roulette schemes mentioned in this text are simple betting strategies that ignore the fact that successive spins of a roulette wheel are independent:

- Watch the game until either low (1–18) or high (19–36) numbers have failed to come up five times in a row, then start betting that half of the layout (the "cold" or "due" numbers fallacy).

- Going in the other direction: wait until three straight odd or even numbers have come up, and then bet *with* the streak (the "hot" numbers fallacy).

- Wait until one of the columns on the layout has not come up for seven spins, then bet that column ("cold" numbers again).

About the only advantage to be derived from strategies like these is that you'll lose, on the average, less money by virtue of not betting on every spin while you wait for the necessary streak to be established. The unnamed author of this brochure does note that these systems do not guarantee a win, but then assures the reader that he has "won consistently" using them.

A betting strategy from *Famous Las Vegas Gambling Systems* that does not rely on disregarding independence calls for repeated betting on one of the columns; these bets pay off at 2 to 1. For an initial bet of $1, the scheme calls for the following:

- On any winning bet $\left(p = \dfrac{12}{38}\right)$, take down your winnings and let the original bet ride.

- If the first bet loses, bet $1 again.

- If the second bet also loses, bet $2 on the third spin.

- Follow this "two losing spins, then double your bet" strategy until you win a bet.

The author then states that "you may do this [double your bet after two straight losses] as many times as you like." This is just the martingale approach with a slight delay built in, as you will double a bet only after two consecutive losses. Of course, the 2 to 1 payoff means that you'll be ahead after a win, but the risk of a long string of losses, during which the amount wagered slowly climbs, is greater than with the martingale on an even-money bet.

6.2 Craps Strategies

Patience

In [104], John Scarne describes the following scam, called the *Watcher* or *Patience System*. The purveyors of this betting strategy either have no understanding of the concept of conditional probability or understand it and are deceiving their customers. Neither of those is a ringing endorsement of their methods.

> *Make $10 at craps almost every time. Simply hold your money until the shooter has made four straight passes, then bet $10 on Don't Pass on the fifth come-out roll. Since the probability of a shooter making 5 consecutive passes is less than 3%, your chance of winning is a whopping 97%!*

Here's what's right about this system:

- The probability of a shooter making five straight points is $(.492)^5 \approx .0288$, which is indeed less than 3%. This is because successive rounds of craps starting with a come-out roll are independent, and so the simple version of the Multiplication Rule (Theorem 2.4.1) can be used.

Here's what's wrong about it:

- The probability we should be considering when we make that Don't Pass bet is not the simple probability "What is the chance of a craps shooter making five consecutive points?" but the conditional probability "What is the chance of a craps shooter making 5 consecutive points, *given that he or she has already made four points?*"

Since the individual rounds are independent, this probability is just the probability of making a single point, which is .492. By betting on Don't Pass, your chance of winning is 47.9%, less than half the advertised 97%.

The system goes on, however, by adopting a martingale strategy should you lose that first $10 bet. If the shooter makes that fifth point, simply make a $20 Don't Pass bet on the sixth point, and so on. While a win will take you back to a $10 profit, a long losing streak will once again bump up against the casino's bet limit.

Like the roulette systems mentioned in *Famous Las Vegas Gambling Systems*, one small redeeming feature of this system is its requirement that you wait until four straight passes have been made before making a bet. This will keep you out of action for long periods of time—the probability of four straight passes is $(.492)^4 \approx .0586$, meaning that you won't be betting on about 94% of all rolls—and thus decrease the rate at which you lose money.

Iron Cross

The *Iron Cross* is a combination of craps bets that illustrates Theorem 3.2.1, on the additivity of expected value, perfectly. Many sources (see [55] for an example) are honest enough to admit that the overall expectation is still negative.

To bet the Iron Cross, a gambler makes the following four simultaneous bets:

- A $5 Field bet. Recall from Example 3.2.12 that this is a bet that the next roll will be 2, 3, 4, 9, 10, 11, or 12. At many casinos, a field bet pays 2 to 1 if a 2 or 12 is rolled; it pays even money otherwise.

- Three Place bets (see page 107)—a bet that the number placed will be rolled before a 7—on the numbers 5, 6, and 8. The 6 and 8 place bets pay off at 7 to 6, and so should be made for $6 to avoid rounding in favor of the casino. The place bet on the 5, which pays off at 7 to 5, should be made for $5.

With $22 at risk, Table 6.1 collects the outcomes for each possible roll. The place bets are only resolved if the number placed or a 7 is rolled, and remain active, or "ride," in casino lingo, when any other number is rolled.

TABLE 6.1: Outcomes for the Iron Cross craps bet

Rolls	Field	Place 5	Place 6	Place 8	Net Win	Prob.
2,12	Wins	Rides	Rides	Rides	$10	2/36
3, 4, 9, 10, 11	Wins	Rides	Rides	Rides	$5	14/36
5	Loses	Wins	Rides	Rides	$2	4/36
6	Loses	Rides	Wins	Rides	$2	5/36
8	Loses	Rides	Rides	Wins	$2	5/36
7	Loses	Loses	Loses	Loses	-$22	6/36

Bets that lose are replaced by the bettor. The surface appeal of this bet is easy to see: one of the four bets wins whenever any number other than a 7 is thrown, and thus the Iron Cross returns money 5/6 of the time.

The expected return on each roll of the dice is then

$$E = (10) \cdot \frac{2}{36} + (5) \cdot \frac{14}{36} + (2) \cdot \frac{14}{36} + (-22) \cdot \frac{6}{36} = -\$\frac{14}{36}.$$

Dividing by the \$22 wagered gives a house edge of 1.76%.

A modification of this strategy, called the *Unbeatable Iron Cross*, disregards both the additivity of expectation and the law of independent trials. The bets are the same, but the bettor is directed not to place his or her first bet until a point is established, thus avoiding the possibility of losing all of the bets on a come-out roll of 7. Bets are replaced as they lose until the point is made, at which point all bets that remain are left to ride until they are resolved and are not replaced after a loss [55].

The inherent assumption behind the Unbeatable Iron Cross is that a 7 is somehow "due" after the point has been made, and so risking further money is unwise. Anyone who understands that successive rolls of the dice are independent will appreciate the flaw in this reasoning: the probability of rolling the dreaded 7 remains at $\frac{1}{6}$, regardless of any or all previous rolls.

6.3 Slot Machine Strategies

There are literally dozens of suggestions for slot machine players about how to choose a machine or how to choose their wager in an effort to improve their chances of winning or the amount of their winnings. Few of these stand up to mathematical scrutiny.

The fact of the matter is this: With the computer technology that runs a slot machine generating thousands of random numbers every second, no amount of "strategizing" once play has begun will have any more than a random effect on the outcome. This includes such practices as playing rapidly or slowly, opting to pull a handle rather than push a button to spin the reels, and playing without a player's club card out of suspicion that the computer is set to give lower payoffs to club members.

It is not widely appreciated that identical-looking machines may have very different payoff odds, due to the internal settings on the computers running the games. While information about the exact payoff percentage of a given machine is difficult to obtain—the proprietary algorithms that generate pseudorandom numbers and map them to reel positions are closely-guarded trade secrets—there are some general principles that are valid. One is that the *hold percentage*—the proportion of the wagered money that is retained by the casino—is generally higher for lower denomination machines, and decreases as the denomination rises. Penny slots may return 88–90% of the money inserted, holding 10–12%, while dollar slot machines might return 95% and hold only 5%, and \$5 slots return 95–97.5% [112]. It must be stressed that these are *long-term* percentages; this does not mean that an individual player will get back 97¢ of every dollar played. Over the course of the machine's life, during which it takes in thousands of dollars, a casino operator can rely on the Law of Large Numbers to confidently count on a certain percentage being held as profit. The return to a single player betting a couple of dollars is far more variable.

Example 6.3.1. On a slot machine with a 94% payback percentage, an initial investment of $100 should, *on the average*, return $94. If these proceeds are played back into the machine, the expected return after the second cycle is $100 \cdot .94^2 = 88.36. If the player continues "reinvesting" all of his or her winnings, the expected holdings after n cycles is $100 \cdot .94^n$—an amount that approaches 0 as n increases. ∎

Short-term fluctuations—in either direction—are likely, but the casino will get its percentage in the long run.

This difference in payback percentages becomes more important when modern video slot machines are considered. The denomination may well be 1¢, but if a machine offers 20 paylines and the chance to risk 9 credits per line, a player making the maximum bet (which is often easily facilitated by pressing a MAX BET button) is putting $1.80 on the line with every spin of the virtual reels. This simple act effectively turns a penny machine into a $1 machine, but without the more favorable payback percentage of the higher denomination. The intangible factor is, as always, the entertainment value of the 1¢ machine. Modern video slot machines often resemble intricate video games with complex bonus games that are an entertainment experience in themselves. The risk is that a player will invest far more money than he or she intended in a gaming session chasing "just one more bonus round." While good advice to the slot player would be to play on the highest denomination machine your average bet per spin matches, there are recreational benefits to lower-denomination machines that confound stark mathematical analysis.

Progressive Slots

A strategy that has some validity concerns *progressive* slot machines. These may be identified by the presence of one or more large prize amounts displayed on a video screen above a single machine or a bank of several machines. A typical progressive slot machine operates by extracting a small amount of each bet placed and adding it to the progressive jackpot, which increases with player activity and is awarded under a particularly rare set of circumstances, such as hitting three jackpot symbols on a designated payline with maximum coins bet.

Naturally, in light of the jackpot amounts—often running into millions of dollars—these prizes are very rarely won. Less obvious is the fact that, in order to finance the big jackpot, progressive slot machines typically pay off at a lower rate than other slot machines—even nonprogressive versions of the same model. One valid conclusion that can be drawn is this:

> *If you're going to play a progressive slot machine, make sure that you bet enough to activate the conditions for winning the big jackpot.*

Whether this is activating all payoff lines or making the maximum bet per line, you need to do it. It is entirely possible to hit the right combination of symbols but lose out on the big prize because you haven't wagered enough money on that payline. If betting the necessary amount to make the progressive jackpot available exceeds your bankroll or your comfort level, then you need to find a different slot machine.

Some slot machines have progressive jackpots that are guaranteed to hit before some fixed jackpot amount is reached. On many such machines, the computer chooses a number at random from the minimum value to which the jackpot resets and the maximum value. For example, a given machine may have a jackpot that starts at $1000 and must hit before the jackpot reaches $5000. The jackpot may be triggered when the accumulated value reaches $3652.24. This value is, of course, not displayed to players. As the jackpot rises with continued play, the triggering value is reached, and the player is then awarded the jackpot. This result can be arranged and displayed on the screen by the computer software.

In the case of multiple machines linked over a network, the player whose wager puts the jackpot over the top receives the jackpot.

As the jackpot gets closer to the "must hit by" number, the game may become a positive expectation wager, although the advantage may be tough to identify and the window of opportunity may be limited. If the jackpot is sufficiently large, such a scenario can attract advantage slot machine players, who may team up to monopolize all networked machines that feed the same jackpot and agree to split their accumulated winnings.

Accumulator Slots

One other avenue for advantage slot play involves *accumulator* or *banking* slot machines. In addition to the standard physical or simulated reels, accumulator slot machines have a separate display where certain symbols are collected, randomly as the machine is played. When the number of symbols reaches a certain threshold, a bonus of some sort—free spins or a cash prize—is awarded.

An accumulator slot machine that inspired some advantage play in the early 2000s was based on S&H Green Stamps, which were trading stamps given out with purchases that could be collected and redeemed for cash or merchandise. As a gambler played the S&H machine, occasionally it would award some virtual stamps, which were displayed in a booklet. When the number of stamps hit 1200 and filled a booklet, players were offered the option of collecting 5 free spins or playing on in pursuit of more stamps and bigger prizes. If a player collected 5 full books, an extra 1000-credit bonus was paid and the 25 free spins began.

A player who abandoned the machine, for whatever reason, before filling a stamp book left their accumulated stamps behind. The next player could begin playing with a head start on getting to 1200. This gave rise to an advantage situation: if a player came across an S&H machine with 1000 stamps already in a book, it might not take a large investment (S&H was frequently a penny machine) to pick up the remaining stamps and trigger 5 free spins. Once those were complete, the advantage player would cash out and move on. The payout may have been small—no one was likely to swoop in and play the machine to a life-changing jackpot—but if there were more than 600 stamps on the screen, the machine was likely in positive-expectation territory.

Other banking machines have made it to casino floors; each one requires careful study of its bonus structure to determine whether or not, and when, an arriving player has an edge [58]. Some more recent accumulators will reset the collection after a certain period of inactivity, making the window of opportunity for an advantage player very narrow. Indeed, advantage players are responsible for the disappearance of many popular and lucrative accumulator machines.

6.4   And One That Does: Lottery Strategies

When we claim that the lottery strategies described in this section work, it's necessary to be clear about what that means. Following the ideas presented here to choose your lottery numbers will *not* change your probability of winning. Any set of, say, six numbers from 1 to 47 has the same probability of winning, whether it is a combination like $\{1, 2, 3, 4, 5, 6\}$ or something more random like $\{7, 20, 25, 35, 43, 45\}$. A life-changing lottery jackpot in Powerball or a similar lottery, when it's won, will be split among all tickets bearing the winning numbers, so your goal in picking numbers is to choose numbers that are unlikely to be chosen by other players. You will still have the same (tiny) probability of winning the jackpot, but if you do win, you will be far less likely to have to share your winnings with another player or players. Forget what you learned in elementary school: *sharing [jackpots] is bad.*

Example 6.4.1. One of the lottery options in the Philippines is the *Grand Lotto 6/55*, a thrice-weekly 6/55 drawing. On October 1, 2022, the winning numbers were 9, 18, 27, 36, 45, and 54—all multiples of 9. The jackpot of 236,091,188.40 Philippine pesos (approximately 4 million US dollars) was split among 433 winning tickets. Rather than a multi-million dollar prize, each winner had to be content with $9300 before taxes [20]. ∎

The reaction to this drawing in the Philippines was swift: a disappointed population and an outcry of accusations of fraud among lottery officials. As we shall see, it is not at all unreasonable that multiple gamblers would pick the same numbers in an r/s lottery, and the combination chosen that day has the same probability of winning as any other combination, including all the ones that look more random than this one.

The basic idea behind the strategy outlined here is very simple: Find out what other people are likely to do, and *don't do that*. For purposes of illustration, we shall consider a Classic Lotto 47 game, where players pick six numbers in the range from 1 to 47. Studies have shown that many lottery players tend to pick their lotto numbers using some combination of the following methods [50]:

1. Choose numbers with personal significance, such as birthdays, anniversaries, or other important dates.

2. Choose numbers that make interesting patterns on the bet slip.

3. Choose numbers toward the center of the bet slip.

4. Choose numbers in arithmetic progression, such as the set {7, 14, 21, 28, 35, 42}. Seven, of course, is considered by many to be a lucky number, and so this progression starts with 7 and counts by 7s.

There is some overlap among these methods, of course—numbers in arithmetic progression often make interesting patterns on the bet slip, for example. Avoiding the numbers that these methods tend to choose is the key to minimizing the likelihood that you will have to share the jackpot in the unlikely event that you win it.

The first criterion is the easiest to avoid. Lottery numbers based on birthdays will not be greater than 31, so a player seeking to pick a unique set of numbers should primarily choose high numbers, with no more than one under 31. In practice, the goal should be to choose a set of numbers with a large sum.

Consider the general r/s lottery (page 46), where players select r numbers in the range from 1 to s. Classic Lotto 47, then, is a 6/47 game. In an r/s lottery, the sum of the numbers in any possible combination will have (see [50]) mean

$$\mu = \frac{r \cdot (s+1)}{2}$$

and standard deviation

$$\sigma = \sqrt{\frac{(s-r) \cdot r \cdot (s+1)}{12}}.$$

Your goal in choosing numbers should be to have a sum higher than about 75% of all possible sums. Since the sums of the r numbers chosen in an r/s lottery are bell-shaped and symmetrically distributed about the mean, some simple properties of the normal distribution can be used here. For a normal distribution with parameters μ and σ, the value separating the bottom 75% of values from the top 25% is

$$x = .6745\sigma + \mu,$$

which we round up to the next integer. In a 6/47 lottery such as Classic Lotto 47, we find that the sum of a player's numbers has mean $\mu = 144$ and SD $\sigma \approx 31.37$. If we choose numbers that add up to 166 or greater, we will have a combination whose sum exceeds 75% of all possible sums, and we will avoid most popular date combinations.

Of course, it is necessary to take the other factors listed above and explained below into account, for while the combination $\{42, 43, 44, 45, 46, 47\}$ may have a high sum, it is actually a relatively popular choice and fails our other tests of suitability [50].

Since lottery bet slips are optically scanned, interesting patterns on the bet slip that are not associated with arithmetic progressions tend to be clusters of adjacent numbers. Figure 6.2 shows a 6/47 ticket with 2 clusters. Once

1	2	3	4	5
6	7	8	9	10
11	12	13	14	15
16	17	18	19	20
21	22	23	24	25
26	27	28	29	30
31	32	33	34	35
36	37	38	39	40
41	42	43	44	45
46	47			

FIGURE 6.2: 6/47 lottery ticket with numbers forming 2 clusters.

again, we look at common practice and then do something else. This factor depends on the exact layout of the betting slip, which in turn may depend on s. A general rule is to avoid too many clusters, which may correspond to arithmetic progressions, and also to avoid too few, which result from bettors making nice-looking patterns.

Example 6.4.2. Referring to the ticket shown in Figure 6.2, the selection $\{1, 8, 15, 22, 29, 36\}$ is an arithmetic progression with no adjacent numbers. A ticket for $\{27, 28, 33, 34, 39, 40\}$ confines all of its numbers to a single cluster. ■

Toward that end, we define the *cluster number* of a wager as the number of adjacent—either edge-adjacent or diagonally adjacent—blocks of numbers. A single number in isolation comprises a cluster, as does any string of bet squares connected at their edges or corners. For an r/s lottery, cluster numbers of 1 or s are to be avoided, as they tend to correlate with smaller overall payouts at the high levels [50].

Edge numbers are numbers that appear on an edge of the bet slip, and their exact number and values depend on the format of the slip, as with clusters. Players trying to be "random" in their selections often choose few numbers near the edges, so if you don't want to share your prize, look to the edge numbers in selecting your combination [50]. From a practical perspective, you should choose a combination of 6 numbers containing least 4 edge numbers—keeping in mind, of course, the need for a suitably large sum and an appropriate cluster number.

Finally, we address arithmetic progressions. A combination such as $\{1, 2, 3, 4, 5, 6\}$ is very simple, very popular, and no more or less likely to win than any other. In no small part, this is because the rule for determining this combination is very simple and easily discovered by many people. The same objection may be raised against a combination like $\{5, 10, 15, 20, 25, 30\}$ or even one like $\{8, 15, 22, 29, 36, 43\}$. To avoid such popular methods of choosing numbers, we define the *arithmetic complexity* of a combination:

Definition 6.4.1. The *arithmetic complexity* (AC) of a set of r numbers is the number of positive differences among all of the numbers in that set, minus $r - 1$.

Example 6.4.3. The set $\{5, 23, 25, 28, 44, 46\}$ chosen by an online random number generator leads to the following positive differences:

$46 - 5 = 41$	$46 - 23 = 23$	$46 - 25 = 21$	$46 - 28 = 18$	$46 - 44 = 2$
$44 - 5 = 39$	$44 - 23 = 21$	$44 - 25 = 19$	$44 - 28 = 16$	
$28 - 5 = 23$	$28 - 23 = 5$	$28 - 25 = 3$		
$25 - 5 = 20$	$25 - 23 = 2$			
$23 - 5 = 18$				

Note that every number is subtracted from every other number in such a way that the difference is positive. There are 12 different differences—2, 21, and 23 are repeated—so the AC of this set is $12 - 5 = 7$. ∎

Example 6.4.4. For the set $\{1, 2, 3, 4, 5, 6\}$, the numbers give rise to the following positive differences:

$6 - 1 = 5$	$6 - 2 = 4$	$6 - 3 = 3$	$6 - 4 = 2$	$6 - 5 = 1$
$5 - 1 = 4$	$5 - 2 = 3$	$5 - 3 = 2$	$5 - 4 = 1$	
$4 - 1 = 3$	$4 - 2 = 2$	$4 - 3 = 1$		
$3 - 1 = 2$	$3 - 2 = 1$			
$2 - 1 = 1$				

We count five different differences, so the AC of this set is $5 - 5 = 0$. ∎

When a set of r different numbers is in an arithmetic progression like those above, the AC will always be 0—this is why we subtract $r - 1$. An AC of 0 suggests that the set in question is not very complex, and that certainly applies to an arithmetic sequence. The maximum AC is

$$\frac{r(r-1)}{2} - (r-1) = \frac{(r-2)(r-1)}{2},$$

when all of the differences are different.

Since lottery players tend toward patterns and simple combinations, the gambler seeking an edge should avoid those and choose more complex sets: anything with an AC of at least 6. This excludes few combinations (in a 6/49 lottery, over 96% of combinations have $AC > 6$), but they are popular combinations [50].

Taking everything together, we have arrived at one possible strategy for choosing numbers that will decrease the likelihood of choosing someone else's numbers and thus having to split the prize if you should win: For a 6/47 lottery, play only combinations fitting all four of the following criteria:

- The sum of the numbers is at least 166.
- The cluster number is between 2 and 5.
- The edge number is at least 4.
- The AC is at least 6.

In addition, any combination that satisfies these requirements should nonetheless be discarded if it has a recognizable pattern—for example, {32, 33, 35, 38, 42, 47}, or is a recent winning combination in this lottery or a neighboring state's lottery, for people are known to favor those combinations as well. In the New York Take 5 drawings of October 27, 2022 (page 49), 52 players chose the winning numbers from the 2:30 P.M. drawing as their numbers for the 10:30 P.M. drawing. When the combination repeated, they split the jackpot 52 ways, taking home only $715.50 apiece instead of the lump sum of $37,206 that would have been paid to a lone winner.

Example 6.4.5. Returning to Example 6.4.1, we see that the Grand Lotto 6/55 winning combination consisting of all of the multiples of 9 has arithmetic complexity 0, making it a poor choice. Additionally, the bet slip for the Grand Lotto 6/55, shown in Figure 6.3, has all 6 of these numbers on a common diagonal, giving this combination a cluster number of 1, and includes only 3 edge numbers.

A good combination for this lottery would have a sum of at least 193; this choice adds up to 189 and so falls just short on that measure as well.　■

Another option when buying a lottery ticket is to let a computer select your numbers randomly, an option called "Quick Pick" or something similar. This makes it easier to buy a ticket and may actually be an option worth using if you're trying not to share a prize—provided that your random selection isn't a collection of possible birthday numbers, for example. If you have the chance to examine your Quick Pick numbers before committing to the ticket, this can be an easy way to risk money on a longshot without much likelihood that your numbers will be chosen, nonrandomly, by another gambler.

Powerball Advantage Play

The Powerball jackpot for the drawing of November 7, 2022 was estimated at $1.9 billion. Even with the probability of winning fixed at 1 chance in

1	11	21	31	41	51
2	12	22	32	42	52
3	13	23	33	43	53
4	14	24	34	44	54
5	15	25	35	45	55
6	16	26	36	46	
7	17	27	37	47	
8	18	28	38	48	
9	19	29	39	49	
10	20	30	40	50	

FIGURE 6.3: Philippines Grand Lotto 6/55 bet slip, with multiples of 9 highlighted. This was the winning combination on October 1, 2022.

292,201,338, it is worthwhile to consider the question of an accumulated jackpot so large as to make a \$2 Powerball ticket a positive-expectation venture.

A simple analysis of Powerball might suggest that when the jackpot exceeds twice the number of tickets, or \$584,402,676, then the expected value of a single ticket is positive. This disregards all prizes except for the jackpot. It also ignores tax implications on the prize, which is a more problematic omission that nonetheless can be accommodated by considering only the after-tax portion of the jackpot. However, this does not account for increased ticket demand; if 292,201,338 people each reasoned this way and bought a ticket with a positive expected value, very few of them would actually make money.

Careful analysis can derive a minimum value for a positive-expectation jackpot. The probability of matching k white numbers on a single Powerball ticket is

$$P(k) = \frac{\binom{5}{k} \cdot \binom{64}{5-k}}{\binom{69}{5}}.$$

This probability must be multiplied by $\frac{1}{26}$ to find the chance that the Powerball is also matched, or by $\frac{25}{26}$ to consider an event where the ticket does not match the Powerball.

Table 6.2 shows the probability of winning each of the available Powerball prizes. "+ P" indicates that the player's ticket matches the Powerball.

Let J be the amount of the Powerball jackpot. The expected value of a \$2 Powerball ticket is then

$$E = J \cdot (3.4223 \times 10^{-9}) - 1.68,$$

TABLE 6.2: Powerball win probabilities

Match	Payoff	Probability
5 + P	Jackpot	3.4223×10^{-9}
5	$1,000,000	8.5557×10^{-8}
4 + P	$50,000	1.0951×10^{-6}
4	$100	2.7378×10^{-5}
3 + P	$100	6.8994×10^{-5}
3	$7	.0017
2 + P	$7	.0014
1 + P	$4	.0109
0 + P	$4	.0261

where every outcome other than winning the top prize, including losing $2, is collected into the –$1.68 term. $E > 0$ provided that $J > \$490,936,628$.

This view of Powerball overlooks the fact that as the jackpot rises, the number of tickets sold also increases, as does the probability of sharing the prize—possibly dividing it below the level where the expectation is positive. In 2010, a paper by Aaron Abrams and Skip Garibaldi [2] addressed the interplay between lottery jackpots and ticket demand, attempting to determine whether any lotto jackpot with a positive expectation could rise to the level of a good investment when the increase in ticket sales that accompanies a rising jackpot was taken into account. Their work led to two inequalities that were necessary to identify a good investment:

- $N < J$: The number of tickets sold, N, should be less than the jackpot J.

- $J > 2J_0$: The jackpot must exceed twice the *jackpot cutoff* J_0, which is the product of the number of possible tickets and the proportion of ticket sales allocated to the jackpot plus lottery expenses.

For the current Powerball rules, approximately 82.13% of sales is allocated to the jackpot and overhead, so

$$J_0 = \left[\binom{69}{5} \cdot 26 \right] \cdot .8213 \approx \$240,000,000.$$

Consequently, Powerball qualifies as a good bet if the jackpot is at least $480 million and the number of tickets sold is less than the jackpot amount.

A good bet is not necessarily a good investment; these conditions merely identify a lottery drawing with a positive expected rate of return. Abrams and Garibaldi went on to claim that no Powerball jackpot would ever achieve that second goal.

6.5   How to Double Your Money

Q: What's the best way to double your money in a casino?
A: Fold it in half and put it back in your wallet!

All kidding aside, consider the following scenario, a variation of one posed in [46]: You have $500 and need $1000, and no less than $1000, urgently. With an eye toward doubling your money, you enter a casino, determined to bet until you either hit your goal or lose everything—reaching $999 is as useless to you as losing all $500. What strategy gives you the best chance of achieving your goal?

One option is the "big bet" approach: wagering your entire bankroll on one even-money bet and hoping for the best. This has the advantage of being quick—one way or the other, your dilemma will be resolved in about a minute.

It might be argued, however, that this method is riskier than it needs to be in staking everything on one bet, and that you might be better off making a bet with a higher risk but higher payoff, so that you will have several chances to win.

Consider the big bet approach at roulette, where you'll stake all $500 on a single even-money bet. Your chance of doubling your money is $18/38 = .4737$.

If, instead, you put your $500 down at a blackjack table, your chance of winning is about .49 under moderately favorable rules and using basic strategy (see Section 5.3). With that in mind, though, it should be noted that it is unwise to risk your entire bankroll at the start of any one blackjack hand, as this will leave you with no money to split pairs or double down, as you should do when your first two cards give you an edge over the dealer and the rules allow you to get more money on the table.

If you make your one big bet on the pass line at craps, you have a .492 chance of winning and reaching your goal. If you bet the don't pass line, your chance is about .493, allowing for a come-out roll of 12 that does not resolve your initial bet. If you are playing at a craps table that allows free odds bets, it may be possible to devise a betting strategy that allows you to risk less money up front, but this puts you in a position where a win on the come-out roll won't get you all the way to the goal of $1000.

Let's look at spreading the risk (or spreading the opportunity) across more bets. Roulette has a number of options at the same HA, so we'll begin there. If you bet $30 on a single number and it hits once, you can cash out for $1050 and walk away regardless of how many bets you've lost earlier. Five hundred dollars allows for 16 such bets—and all you need to do is win one. The chance of that happening is

$$P(\text{Win at least once}) = 1 - P(\text{Lose 16 out of 16}) = 1 - \left(\frac{37}{38}\right)^{16} \approx .3473,$$

about 12.5% *less* than making a single even-money bet.

By switching to a street bet on three numbers, your money won't last for quite as many spins, but your chance of winning on any spin is tripled. A $90 bet will give you five chances at a $990 payoff, and since your bet is returned with your win, your total after a single win will exceed $1000. The chance of winning one of those 5 bets is

$$P(\text{Win at least once}) = 1 - P(\text{Lose 5 out of 5}) = 1 - \left(\frac{35}{38}\right)^5 \approx .3371,$$

which is not as good as any previous option.

A trend seems to be developing here. Let's look at the other choices for roulette bets, which are compiled in Table 6.3. In each case, the amount of the bet has been rounded to a convenient number.

TABLE 6.3: Roulette bets in a quest to double $500 to $1000

Bet	Payoff	# of bets	Wager	P(Win at least 1)
Straight	35 to 1	16	$30	.3473
Split	17 to 1	8	$60	.3511
Street	11 to 1	5	$90	.3371
Corner	8 to 1	4	$125	.3591
Basket	6 to 1	3	$166	.3451
Double street	5 to 1	3	$166	.4028
Dozen	2 to 1	1	$500	.3684
Even-money	1 to 1	1	$500	.4737

One note: If you're going to make a dozen bet, your first bet can be for only $250, since a win there gives you $500 in winnings and the return of your wager. Taken together with the $250 you didn't bet, you have $1000. However, if your first bet loses, you then are faced with raising $1000 starting with only $250—a somewhat more daunting task than the original.

The conclusion is clear: If you're trying to double your money at roulette, your best chance to do so is by making a single even-money bet.

The reason for this is very simple and has nothing to do with roulette: The more bets you make, the greater the opportunity for the house advantage, whatever the game or the edge may be, to work against you. The phrase "Go big or go home" is highly appropriate to this quest, and this applies regardless of the game you're playing. If your tastes run more toward slot machines, you're better off making your bets on the highest denomination machine you can find, because the rate of return is typically higher on higher-denomination slot machines, and fewer spins will be necessary to win the money you seek. Find your way to the high-limit room of your favorite casino. If there's no $500 machine available (they exist, and pay out as much as 1000 to 1 on certain reel combinations) then you should seek out the highest denomination they do provide.

Example 6.5.1. A bet with multiple payoffs, such as the Field bet at craps, might suggest an alternate strategy, since the possibility of a better-than-even payoff provides the option of making a smaller initial bet and holding back some money for a second wager. If you bet \$250 on the first roll, there's a $\dfrac{1}{18}$ chance that a 2 or 12 will be rolled, bringing your total immediately to \$1000; if you lose that bet, you still have \$250 for another chance at reaching \$1000.

Suppose that you bet \$250 on each roll until you have at least \$1000 or run out of money. Let E be the event that you reach or exceed your \$1000 goal, and define $\hat{p} = P(E)$. The following disjoint events are part of E:

- E_1: A 2 or 12 is rolled on the first roll, and you have \$1000. The probability of this event is

$$p_1 = \frac{2}{36}.$$

- E_2: You win even money on two consecutive rolls, bringing your total to \$1000. This event has probability

$$p_2 = \left(\frac{14}{36}\right)^2 = \frac{196}{1296}.$$

- E_3: You win even money on the first roll and 2 to 1 on the second roll, for a total of \$1250.

$$P(E_3) = p_3 = \frac{14}{36} \cdot \frac{2}{36} = \frac{28}{1296}.$$

- E_4: You lose the first roll, win 2 to 1 on the second, and win either 1 to 1 or 2 to 1 on the third roll. Your total is either \$1000 or \$1250, and the probability of this event is

$$p_4 = \frac{20}{36} \cdot \frac{2}{36} \cdot \frac{16}{36} = \frac{640}{46,656}.$$

- E_5: In the first two rolls, you win one and lose one, returning you to a balance of \$500. When your balance is \$500, your probability of eventually winning is $\hat{p}$. Since there are two different orders (win-lose and lose-win), we have

$$p_5 = 2 \cdot \frac{14}{36} \cdot \frac{20}{36} \cdot \hat{p} = \frac{560}{1296} \cdot \hat{p}.$$

- E_6: You lose on the first roll, win 2 to 1 on the second to bring your total to \$750, and lose the third roll to return you to \$500, from which your chance of winning is again $\hat{p}$.

$$P(E_6) = p_6 = \frac{20}{36} \cdot \frac{2}{36} \cdot \frac{20}{36} \cdot \hat{p} = \frac{800}{46,656} \cdot \hat{p}.$$

Adding everything up gives

$$\hat{p} = p_1 + p_2 + p_3 + p_4 + p_5 + p_6 = \frac{353}{1458} + \frac{655}{1458} \cdot \hat{p},$$

which is a linear equation in $\hat{p}$ whose solution is

$$\hat{p} = \frac{353}{803} \approx .4396.$$

∎

This is still inferior to the probability of .492 of doubling your money by making a single pass line wager.

6.6 Exercises

Answers to starred exercises begin on page 290.

6.1. * More recently, the roulette system published in *Bohemia* (Example 6.1.1) has been called the "Three-Two System," and calls for a bet of 3 units on black and 2 units on the third column [146]. Find the expectation of this version of the bet.

6.2. The *d'Alembert* system for roulette is said to be the work of French mathematician Jean le Rond d'Alembert. Its fundamental premise is an excellent illustration of the Gambler's Fallacy. The description here (from [4]) assumes that you're betting on red:

> *Since red and black occur equally often, you simply increase your bet by one unit after each loss and decrease it by one unit after each win until you get back to your original starting bet, and then bet one unit each spin until you lose.*

Apart from its embrace of incorrect mathematics in its disregard for independence, find the flaw in this system.

6.3. Here's a roulette system posted on the Internet in July 2012:

> *Beginning with a bankroll of $450, wager on a single number until either it wins or you go broke after 50 straight losses. On spins 1–20, wager $5 per spin. On spins 21–40, wager $10 per spin. On spins 41–50, wager $15 per spin.*
> *As soon as your number hits, you quit and leave with a profit.*

Assume that you're playing American roulette.

a.* Find the probability that your number will hit exactly once in 50 spins.

b. By computing your total holdings if your number comes up for the first time on spins 20, 40, and 50, show that you will indeed be ahead if your number hits once.

c.* Find the probability that you will lose all of your initial bankroll.

6.4. Absent the potential for a 2 to 1 payoff when a 2 or 12 is rolled, doubling your money on a Field bet would be a simple "one-shot" problem. Show that the probability that you will double your money from $500 to $1000 on a single Field bet is greater than the probability of .4396 calculated in Example 6.5.1.

6.5. The *Red Snake* strategy at roulette has been imbued by some with mystical powers. Like the *Bohemia* magazine strategy described in Example 6.1.1, it relies on a pattern in the roulette layout. To make the Red Snake bet, a gambler wagers on the 12 red numbers 1, 5, 9, 12, 14, 16, 19, 23, 27, 30, 32, and 34, which form a continuous string of red numbers crossing back and forth across the table. Show that the house advantage of the Red Snake wager on an American roulette wheel is the same 5.26% that most other roulette bets face.

6.6. In [88], we find the following roulette system, which relies on making single-number bets and hoping for a big payoff:

> *There are 38 numbers on the wheel. One number, played nineteen times, has literally a fifty percent chance of showing; that's even odds! But you don't play that way for even money; if and when your number shows, you are paid 35 to 1!*

We've seen this error before—the author is making the same miscalculation that the Chevalier de Méré made. Find the correct probability that a single given number will turn up at least once in 19 spins of an American roulette wheel.

6.7.* In the scenario described in Exercise 6.6, suppose that you choose to bet on the number 00. The number of 00s in 19 spins is a binomial random variable. Calculate the expectation of 19 straight bets on 00.

6.8. The *Shotwell* system for roulette betting, described in [123], relies on the proximity of certain numbers to one another on an American roulette wheel, which is shown in Figure 6.4.

The Shotwell system combines one double street (six-number) bet with four specific single-number bets, chosen so that the ten numbers covered by the bet are no more than three spaces apart on the wheel. The choices for the Shotwell system are given in Table 6.4.

The author of this system asserts that every time the wheel is spun, you have a 1 in 4 chance of winning, and closes by saying "Enjoy your winnings—the casino won't."[123]. Show that the house edge for this bet is still 5.26%.

FIGURE 6.4: American roulette wheel layout [29].

TABLE 6.4: Betting options for the Shotwell roulette system

Six-number bet	Single-number bets
1–6	8, 10, 20, 26
4–9	10, 13, 14, 15
10–15	16, 17, 18, 28
19–24	1, 2, 4, 26
28–33	00, 22, 24, 35
31–36	0, 00, 29, 30

6.9. The Shotwell system goes on to advocate increasing your bets after a win. After a single-number bet hits, the system calls for doubling all of the bets until they lose—an act that does not change the HA. If the double street bet wins, the next bet should be double the previous bet on the double street, while the four single-number bets are unchanged. Confirm that the house edge remains 5.26% in this latter scenario.

6.10. In [131], we find the following craps system, purported to be a way to "risk \$2700 to win \$25": The bet is that four consecutive shooters will *not* make at least four straight passes—that is, that somewhere in four shooters, one of them will fail to make a point among their first four decisions.

Here are the details, with all wagers on Don't Pass: For shooter #1, begin with a \$25 bet. If the shooter fails to make the point, you win \$25. If the shooter makes his or her point, bet \$50 on the next come-out roll. If this point is not made, you win \$50, for a net profit of \$25. If the shooter makes the second point, bet $100 on the next come-out roll, and if that point is made, bet $200 on the fourth roll. If the shooter makes four straight points, stop betting until the streak ends and a new shooter takes the dice. Your current loss is $375.

Repeat this action on the second shooter, but with successive wagers of $50, $75, $150, and $300. Again, once a point is missed or the shooter rolls a 2 or 3 on the come-out roll, collect your $25 profit and begin again at the start of the sequence. If the second shooter makes four straight passes, wait for the third shooter and begin betting with the sequence $75, $100, $200, $400. If shooter #3 makes four straight passes, start the sequence $100, $125, $250, $500 with the fourth shooter.

If all four shooters make at least four passes, your total loss is $2700. If not, you've made at least a $25 profit with your last bet.

a. Using the probabilities on page 23, find the probability of losing the first four bets.

b. Find the probability of losing 16 straight bets and hence the full $2700.

c.* If you lose the first four bets and are down $375, find the probability that the next 15 shooters will each fail to make four straight passes, thus allowing you to recoup all of your losses.

6.11. Given the following bet slip for a 6/47 lottery, construct a "good" combination of numbers that meets all of the criteria in Section 6.4:

1	2	3	4	5
6	7	8	9	10
11	12	13	14	15
16	17	18	19	20
21	22	23	24	25
26	27	28	29	30
31	32	33	34	35
36	37	38	39	40
41	42	43	44	45
46	47			

Appendix A: House Advantages

Wager	HA	Page
Blackjack, single-deck, Las Vegas Strip rules	~ 0.00%	230
Craps, pass line bet with 1000X odds	.0014%	175
Craps, pass line bet with 4X odds	0.28%	174
Super Bowl coin toss proposition bet	0.98%	144
Baccarat, Banker bet with 5% commission	1.06%	119
Baccarat, Player bet	1.23%	119
Craps, don't pass or don't come line	1.34%	103
Craps, pass or come line	1.41%	103
European roulette, all bets	2.70%	100
Three card poker, Q64 as beacon hand	3.37%	140
Three card poker, call on queen or higher	3.45%	140
American Royal Roulette, all bets	4.00%	167
American roulette, Colors bet	4.34%	166
Sports betting, one game	4.55%	141
American roulette, all bets except basket bet	5.26%	99
Crapless craps, pass line bet	5.40%	173
Craps, field bet	5.56%	105
Chuck-a-luck	7.87%	102
American roulette, basket bet	7.89%	101
Craps, hardway bet on 6 or 8	9.09%	107
Blackjack, Royal Match side bet	10.90%	222
Big Six wheel, bet on $1	11.11%	102
Craps, hardway bet on 4 or 10	11.11%	107
Baccarat, Tie bet	14.05%	119
Sic bo, bet on 4 or 17 (correct payoff)	15.28%	184
Double Action Roulette, single number, both wheels	16.83%	168
Craps, Any Seven bet	16.67%	104
Keno, FireKeepers Casino 20-spot game	17.01%	111
Policy game, day number bet	23.08%	115
Craps, Fire Bet	24.90%	176
Keno, Mark 7 bet	36.30%	109
Michigan State Lottery Daily 3 straight bet	50.00%	112
Michigan State Lottery Daily 3 box bet	50.20%	112

DOI: 10.1201/9781003358183-A

Answers to Selected Exercises

Chapter 1

Exercises begin on page 29.

1.1a. $9/38$.

1.1b. $9/38$.

1.2. 3.075×10^{-4}.

1.3. $1/17$.

1.4. For a royal flush, there are only four choices—the suit—hence four possible hands. For Benny Hill's hand of four aces and a king, the only variable is the suit of the king, and again there are four choices. In either case, there are 4 possible hands, and so the probabilities of the 2 hands are identical—we don't need to know how many 5-card poker hands are possible.

1.5. Highest rank: four-of-a-kind—for example, $JJJJK$. Lowest rank: 2 pairs, such as $QQJJK$.

1.6. $P(\text{Win an Over 7 bet}) \approx .4167$. By symmetry, this is also the probability of winning the Under 7 bet.

1.7. $P(\text{Win}) = .15625, P(\text{Lose}) = .0625$.

1.9. With 16 barred, $P(\text{Win}) \approx .5184 > .5$. With 2 and 16 barred, $P(\text{Win}) \approx .5027 > .5$.

Chapter 2

Exercises begin on page 86.

2.3. $.0506$.

2.4. $.0211$.

2.7. Let r be the number of matched red numbers, w the number of matched white numbers, and $P(r, w)$ the probability of matching r red and w white numbers. The following table gives the values of $P(r, w)$.

r	w	$P(r,w)$
2	2	$1/106,525$
2	1	$48/106,525$
1	2	$48/106,525$
2	0	$276/106,525$
0	2	$276/106,525$
1	1	$2304/106,525$
1	0	$13,248/106,525$
0	1	$13,248/106,525$

2.8. Matching 3 numbers plus the Powerball is (slightly) more likely.
2.10. 6.4×10^{-11}.
2.12. 190.
2.13. .1961.
2.14a. $P(15) \approx 2.3951 \times 10^{-8}$.
2.14b.

$$P(16) \approx 6.6828 \times 10^{-10}.$$
$$P(17) \approx 1.1035 \times 10^{-11}.$$
$$P(18) \approx 9.5125 \times 10^{-13}.$$
$$P(19) \approx 3.3943 \times 10^{-16}.$$
$$P(20) \approx 2.8286 \times 10^{-19}.$$

2.14c. 2.4631×10^{-8}.
2.16. 12.10%.
2.17. .3679.
2.18. .0024.
2.19. .0103.
2.20. $P(A) \approx .1151; P(B) \approx .1094$, so A is more likely.
2.22. .0455.
2.24. .0353.
2.25. The most likely of the 3 events is 56 straight passes, with probability

$$6.2509 \times 10^{-18} \approx \frac{1}{160,000,000,000,000,000}.$$

Fourteen straight 14s at American roulette has probability 7.6389×10^{-23}, approximately double the 3.3287×10^{-23} probability of 17 straight naturals.
2.27. $p = \dfrac{38!}{38^{38}} \approx 4.8612 \times 10^{-16}$.

Chapter 3

Exercises begin on page 157.

3.1a. $P(7) = \frac{1}{6}$.

b. $\frac{4}{36} = \frac{1}{9}$.

c. Careful counting of the possibilities will reveal that the probability distribution of the sum of a pair of Sicherman dice is the same as that for a pair of standard d6s.

3.3. $E \approx -.1111$. The HA is 11.11%.

3.5. The HA is 4.00%, rather higher than the HA of 1.52% for placing the 6.

3.6. The probability of winning falls, since 2 winning rolls have been converted to pushes while only 1 pushing roll is now a winner. The new expected value is $-.0414$.

3.7a. For the straight bet, the expectation is $-\$.50$, and the HA is therefore 50%.

3.7b. For an n-way boxed bet with net winnings of $\$x$, we have

n	x	E	**HA**
24	207	$-\$.5008$	50.08%
12	415	$-\$.5008$	50.08%
6	833	$-\$.4996$	49.96%
4	1249	$-\$.5000$	50.00%

3.8a. $\left(\dfrac{1}{38}\right)^2 = \dfrac{1}{1444}$.

3.8b. $38\dfrac{8}{9} - 1$.

3.8c. 2.98%.

3.9. 29.67%.

3.10. $\dfrac{1}{1001}$.

3.11. $-.7640$.

3.12. $P(\text{Win}) = \dfrac{\dbinom{75}{n-3}}{\dbinom{78}{n}} = \dfrac{n \cdot (n-1) \cdot (n-2)}{456,456}$.

3.13a. .0058.

3.13b. 2.69×10^{-4}.

3.13c. 19.95%.

3.14. 22.22%.

3.15. The HA drops from 35.18% to 18.65%; nearly a 50% reduction.

3.16. The HA of 31.62% is slightly better for players than the standard payoff in Nebraska, but inferior to the Quarter Mania edge.

3.17. No five-card poker hand can contain both a full house and a four-card royal flush.

3.19. For a bet on the $2 spot, $E \approx -\$.1667$.

For a bet on the $5 spot, $E \approx -\$.2222$.

For a bet on $10, $E \approx -\$.1852$.

For a bet on $20, $E \approx -\$.2222$.

For a bet on either one of the two logos, $E \approx -\$.2407$.

3.20. $p > \sqrt[n]{\dfrac{b}{a+b}}$.

3.23. $-\$.01625$.

3.25. $.2461$.

3.27. $-\$.125$, and so the HA is 12.5%.

3.29. 10.94%.

3.31. $-\$.1078$, so the house advantage is 10.78%.

3.32a. $.6976$. This includes both rolling a 7 before the point and 3-roll sequences that result in a push.

3.32b. $.3023$.

3.32c. 6.05%.

3.34. 3.55%.

3.35. $-\dfrac{1}{3}$.

Chapter 4

Exercises begin on page 201.

4.1. 11.52%.

4.2a. 2.77%.

4.2b. 160.

4.3a. 30.6%.

4.3b. 25%.

4.4. 5.26%.

4.6a. 0. (Yes, this bet is fair!)

4.6b. $-\$.08$.

4.6c. Bet only on Blue.

4.8. $-.0540, -.0789$.

4.9. 14.3% for either bet.

4.10. $P(\text{Win}) = {}^5/_{18}$. $E = -{}^1/_{18}$.

4.11. Yes. The expected value of a bet is positive.

4.13. 36.1%.

4.15. $E \approx -.0278$.

4.17. Jacks or Better.

4.18. $K\spadesuit\ K\diamondsuit\ K\clubsuit\ T\clubsuit\ 9\heartsuit$.

4.19. 60.

4.20. Careful counting will show that a royal flush is an unbeatable Omaha hand, due to the requirement that a player's final hand use exactly 2 of his or her hole cards. The probability is 0. (In Texas hold'em, the only way for two or more players to tie with a royal flush is if all 5 of the cards appear on the board, but a royal flush on the board in Omaha cannot be completely used by any player.)

4.21. $52.78.

4.22. 7.89%.

4.23. 2.38%.

4.24a. 80.93%.

4.24b. 54.32%.

Chapter 5

Exercises begin on page 252.

5.1. The true count is −4. You should not make the insurance bet.

5.3. 2.845×10^{-5}.

5.5. .896%.

5.9. $1/782,382$. In the infinite deck approximation, $p \approx 2.188 \times 10^{-6}$.

5.10. $1/456,976 \approx 2.1883 \times 10^{-6}$.

5.13. .0270.

5.15a. 8.92%.

5.15b. Approximately 11.08.

5.16. As a function of the number of decks n, the probability of winning is

$$P(n) = \frac{2n}{52n - 1}$$

at the top of a fresh shoe. For $n = 6$, we have $P(\text{Win}) = 12/311 \approx .0386$.

5.17. Approximately $1/391,191$.

5.18.

Event	Probability
4 aces	2.7438×10^{-5}
3 suited aces	5.9492×10^{-5}
3 aces	1.4457×10^{-3}
2 suited aces	6.4028×10^{-3}
2 aces	.0231
1 $A\Diamond$	.0610
1 ace	.1831

Chapter 6

Exercises begin on page 279.

6.1. $-5/19$.

6.3a. .3562.

6.3c. .2636.

6.7. $-$1$.

6.10a. .086.

6.10b. 1.17×10^{-5}.

6.10c. .4042.

Bibliography

[1] 2by2—Prizes and Odds. Online at http://www.powerball.com/2by2/2by2_prizes.asp. Accessed 12 September 2022.

[2] Abrams, Aaron, and Skip Garibaldi, Finding Good Bets in the Lottery, and Why You Shouldn't Take Them. *The American Mathematical Monthly* **117**, January 2010, pp. 3–26.

[3] Andrews, Chris, *Then One Day...: 40 Years of Bookmaking in Nevada*. Huntington Press, Las Vegas, 2019.

[4] Anonymous, *Casino Confidential*. Quirk Books, Philadelphia, PA, 2008.

[5] Baldwin, Roger R., Wilbert E. Cantey, Herbert Maisel, and James P. McDermott, The Optimum Strategy In Blackjack. *Journal of the American Statistical Association* **51**, September 1956, pp. 429–439.

[6] Bernstein, Peter L., *Against The Gods: The Remarkable Story of Risk*. John Wiley & Sons, New York, NY, 1998.

[7] Blackbridge, John, *The complete poker-player: a practical guide book to the American national game: containing mathematical and experimental analyses of the probabilities at draw poker*. Dick & Fitzgerald, New York, NY, 1880.

[8] Bollman, Mark, *Mathematics of Casino Carnival Games*. CRC Press/Taylor & Francis, Boca Raton, FL, 2020.

[9] Bollman, Mark, *Mathematics of Keno and Lotteries*. CRC Press/Taylor & Francis, Boca Raton, FL, 2018.

[10] Bollman, Mark, *Mathematics of the Big Four Casino Table Games*. CRC Press/Taylor & Francis, Boca Raton, FL, 2021.

[11] The Bone Man, *New Craps Fire Bet Emerging in Las Vegas*. Online at http://www.nextshooter.com/firebet. Accessed 23 June 2012.

[12] Brokopp, John, *Gaming Company Makes Roulette More Colorful*. Online at http://brokopp.casinocitytimes.com/article/gaming-company-makes-roulette-more-colorful-58988. Accessed 21 May 2012.

[13] Buro, Chris, Atlantic City Update. *Blackjack Forum* **XVI** #4, Winter 1996, pp. 49–52.

[14] Catlin, Donald E., A Really Hard Hardway Bet, in *Finding the Edge*, Olaf Vancura, Judy A. Cornelius, and William R. Eadington, editors. Institute for the Study of Gambling and Commercial Gaming, Reno, NV, 2000, pp. 297–302.

[15] Chavez, Will, Ball-and-dice games open at Hard Rock Tulsa, *Cherokee Phoenix*, 23 August 2018. Online at https://www.cherokeephoenix. org/multimedia/ball-and-dice-games-open-at-hard-rock-tulsa/article_ ee3cfca5-d15b-5f10-9093-0cb4c5a10ddb.html. Accessed 26 October 2022.

[16] Cheung, Y.L., Why poker is played with five cards. *The Mathematical Gazette* **73** #466, December 1989, pp. 313–315. Online at http://www.jstor.org/stable/3619303. Accessed 12 March 2021.

[17] Devlin, Keith, *The Unfinished Game*. Basic Books, New York, NY, 2008.

[18] Dick, William B., *The American card player, containing clear and comprehensive directions for playing the games of euchre, whist, bezique, all-fours, pitch, commercial pitch, French fours, all fives, casino, cribbage, straight and draw poker, and whiskey poker; together with all the laws of those games*. Dick & Fitzgerald, New York, NY, 1866.

[19] Egen, Captain Frederick W., *Plainclothesman: A Handbook of Vice and Gambling Investigation*. Arco Publishing Company, Inc., New York, NY, 1959.

[20] Elemia, Camille, and Mike Ives, 433 People Won a $4 Million Lottery. Was It Pure Luck?. *The New York Times*, 5 October 2022Online at https://www.nytimes.com/2022/10/05/world/asia/philippines-lottery-jackpot.html. Accessed 9 October 2022.

[21] Emert, John, and Dale Umbach, *Inconsistencies of "Wild-Card" Poker*, CHANCE, 9, 1996, pp. 17–22.

[22] Epstein, Richard A., *The Theory of Gambling and Statistical Logic*, revised edition. Academic Press, San Diego, CA, 1995.

[23] Epstein, Richard A., *The Theory of Gambling and Statistical Logic*, second edition. Academic Press, San Diego, CA, 2013.

[24] Ethier, S. N., & Hoppe, F. M. Teaching a University Course on the Mathematics of Gambling. *UNLV Gaming Research & Review Journal*, **24**(1), 2020. Retrieved from https://digitalscholarship.unlv.edu/grrj/vol24/iss1/2. Accessed 20 May 2021.

[25] *EZ Baccarat: Game Math.* Online at http://www.deq.com/en/table-games/ez-baccarat-math.php. Accessed 3 September 2012.

[26] *Famous Las Vegas Gambling Systems*, Castindes Books Tijuana, n.d.

[27] Ferguson, Tom W., *The Curse of the Big Lottery Win.* Online at http://www.lastchaser.com/BugleTwo.html#anchor_141, 30 September 2012. Accessed 4 October 2012.

[28] Film8ker, American roulette table layout.gif. Online at http://commons.wikimedia.org/wiki/File%3AAmerican_roulette_table_layout.gif. Accessed 16 May 2013.

[29] Film8ker, American roulette wheel layout.gif. Online at http://commons.wikimedia.org/wiki/File:American_roulette_wheel_layout.gif. Accessed 26 July 2013.

[30] Fomin, Dmitri, English pattern playing cards deck.svg. Online at https://commons.wikimedia.org/wiki/File:English_pattern_playing_cards_deck.svg. Accessed 14 April 2021.

[31] The Football Pools. Online at http://www.footballpools.com/cust. Accessed 18 June 2012.

[32] Frere, Thomas, *Hoyle's Games, Containing all the Methods of Playing the Latest and Most Fashionable Games.* J.S. Locke & Co., Boston, MA, 1876.

[33] Frey, Richard L., *According to Hoyle.* Fawcett Crest Books, Greenwich, CT, 1970.

[34] Frome, Elliot, It All Ends with a Razgu. *Gambatria*, 23 January 2014. Online at http://www.gambatria.com/blog/2014/1/23/it-all-ends-with-a-razgu. Accessed 6 March 2021.

[35] Frome, Elliot A., *Expert Strategy for Blackjack Switch.* Gambatria, Las Vegas, 2011.

[36] Galilei, Galileo, *Sopa le Scoperte Dei Dadi.* Online at http://www.leidenuniv.nl/fsw/verduim/stathist/galileo.htm. Accessed 5 September 2013.

[37] Gaming Today staff, '7 Point 7' livens up craps. *Slots Today* **33** #42, October 2008, 14–20, pp. 1–2.

[38] Gelman, Andrew, and Deborah Nolan, You Can Load a Die, But You Can't Bias a Coin. *The American Statistician* **56**, #4, November 2002, pp. 308–311.

[39] Gordon, Dan, How to Bet NFL Futures. *Blackjack Forum* **XXI** #2, Summer 2001, pp. 7–21.

[40] Gordon, Phil, and Jonathan Grotenstein. *Poker: The Real Deal.* Simon Spotlight Entertainment, New York, NY, 2004.

[41] Griffin, Peter, *Extra Stuff: Gambling Ramblings.* Huntington Press, Las Vegas, 1991.

[42] Griffin, Peter, *The Theory of Blackjack*, 6th edition. Huntington Press, Las Vegas, 1999.

[43] Gurtzner, Charles, Dimes Make Millions for Numbers Racket; 600–1 Payoff Lures 500,000 a Day to Make Bets Here. *The New York Times*, 26 June 1964, p. 1.

[44] H., B., Around the World: Canada. *Blackjack Forum* **XXII** #2, Summer 2002, pp. 63–65.

[45] Hadsall, Joe, *No Dice: Casino Invents Version of Craps Played with Cards.* Joplin *Globe*, Joplin, MO, 16 April 2010. Online at http://www.joplinglobe.com/enjoy/x1687713220/No-dice-Casino-invents-version-of-craps-played-with-cards. Accessed 24 June 2012.

[46] Haigh, John, *Taking Chances: Winning with Probability*, 2nd edition. Oxford University Press, Oxford, 2003.

[47] Hannum, Robert C., and Anthony N. Cabot, *Practical Casino Math*, 2nd edition. Institute for the Study of Gambling and Commercial Gaming, Reno, NV, 2005.

[48] Hand, David J., Never Say Never. *Scientific American* **310** #2, February 2014, pp. 72–75.

[49] Hawkins, Jim, Around the States: Mississippi. *Blackjack Forum* **XV** #1, March 1995, pp. 67–68.

[50] Henze, Norbert, and Hans Riedwyl, *How to Win More: Strategies for Intcreasing a Lottery Win.* A K Peters Ltd., Natick, MA, 1998.

[51] Holloway, Louis G, Traveling gamblers: Stick to Nevada. *Systems & Methods* **10**, 1976, p. 43.

[52] Horne, Andrew, File:Michigan Stadium Sept 17,2011.jpg. Online at https://commons.wikimedia.org/wiki/File:Michigan_Stadium_Sept_17,2011.jpg. Accessed 16 October 2022.

[53] How, Stephen, *Card Craps (+EV).* Online at http://discountgambling.net/playcraps/. Accessed 24 June 2012.

[54] Humble, Lance, and Carl Cooper, *The World's Greatest Blackjack Book.* Doubleday and Company, Garden City, NY, 1980.

[55] *The Iron Cross Craps System*. Online at http://www.crapscasinos.org/craps-systems/the-iron-cross/. Accessed 17 October 2012.

[56] Jacobson, Eliot, *Card Counting in Baccarat*. Online at http://www.worldgameprotection.com/the-catwalk/casino-ology/BOSVIEW/Card-Counting-in-Baccarat/. Accessed 24 July 2012.

[57] Jacobson, Eliot, *Contemporary Casino Table Game Design*. Blue Point Books, Santa Barbara, CA, 2010.

[58] Jensen, Marten, *Beat The Slots*. Cardoza Publishing, New York, NY, 2002.

[59] Jensen, Marten, *Video Poker and Slots for the Winner*. Cardoza Publishing, New York, NY, 2002.

[60] Jimpoz, Craps layout.png. Online at http://en.wikipedia.org/wiki/File:Crapslayout.png. Accessed 29 October 2022.

[61] Kaplan, Lawrence J., and James M. Maher, The Economics of the Numbers Game. *The American Journal of Economics and Sociology* **29**, # 4 (October 1970, pp. 391–408.

[62] Keller, John W., *The Game of Draw Poker*. White, Stokes, & Allen, New York, NY, 1887.

[63] Kolata, Gina, In Shuffling Cards, Seven is Winning Number, *New York Times*, 1990.

[64] Krigman, Al, *Beware of Rule Variations: Little Things Can Mean a Lot*. Online at http://krigman.casinocitytimes.com/article/beware-of-rule-variations-little-things-can-mean-a-lot-60995, 3 September 2012.

[65] *The Kryptos System for Roulette*, in *The Big Winner's Systems Book*. Gambling Times, Inc., Hollywood, CA, 1981.

[66] Lampe, Kristen A., Reconciling Wild Cards and Poker. *CHANCE*, **24** #4, 2011, pp. 17–23.

[67] Las Vegas Advisor, Letters and Tips. *Las Vegas Advisor*, April 1994, p. 11. Online at https://www.lasvegasadvisor.com/vegas-with-an-edge/wp-content/uploads/2023/01/100LVA.pdf.

[68] Las Vegas Advisor, *Question of the Day April 7, 2009*. Online at http://www.lasvegasadvisor.com/qod.cfm?qid=1656. Accessed 17 July 2012.

[69] Las Vegas Advisor, *Question of the Day 16 May 2008*. Online at https://www.lasvegasadvisor.com/question/vgk-futures-bets/. Accessed 14 December 2022.

[70] Levinson, Horace C., *Your Chance To Win*. Farrar and Rinehart, Inc., New York, NY, 1939.

[71] Liddick, Don, *The Mob's Daily Number: Organized Crime and the Numbers Gambling Industry*. University Press of America, Inc., Lanham, MD, 2020.

[72] Live Casino Comparer, Evolution Dreamcatcher. Online at https://www.livecasinocomparer.com/live-casino-software/evolution-live-casino-software/evolution-dreamcatcher/. Accessed 26 October 20228.

[73] Lotto Alaska, How To Play Chase the Ace Lottery. Online at https://lottoalaska.com/. Accessed 26 October 2022.

[74] Louw, Stephen, African numbers games and gambler motivation: 'Fahfee' in contemporary South Africa. *African Affairs* **117**, January 2018, pp. 1–21.

[75] Mackenzie, Alsidair, How do the Football Pools work?. 10 June 2022. Online at https://www.fourfourtwo.com/features/how-do-the-football-pools-work. Accessed 11 November 2022.

[76] Mai, Peter, Video Poker Machines.jpg. CC BY-SA 2.0 https://creativecommons.org/licenses/by-sa/2.0, via Wikimedia Commons. Online at https://commons.wikimedia.org/wiki/File:Video_Poker_Machines.jpg. Accessed 5 March 2021.

[77] *Master Library*. Texas Instruments, Inc., Dallas, TX, 1979.

[78] Meadow, Barry, *Blackjack Autumn*. Huntington Press, Las Vegas, 1999.

[79] Mezrich, Ben, *Bringing Down the House: The Inside Story of Six M.I.T. Students Who Took Vegas for Millions*. The Free Press, New York, NY, 2002.

[80] Mezrich, Ben, *Busting Vegas: The MIT Whiz Kid Who Brought the Casinos to Their Knees*. William Morrow, New York, NY, 2005.

[81] Michigan State Lottery, *Where The Money Goes*. Online at https://www.michiganlottery.com/about/where-the-money-goes. Accessed 12 September 2022.

[82] Nebraska Lottery, Nebraska MyDaY. Online at https://www.nelottery.com/homeapp/lotto/33/gamedetail. Accessed 22 October 2022.

[83] Nevada Gaming Control Board, Approved Gambling Games Effective August 1, 2021. Online at https://gaming.nv.gov/Modules/ShowDocument.aspx?documentid=7097. Accessed 4 November 2022.

[84] Nazox, File:Keno goose.jpg. Online at https://commons.wikimedia. org/wiki/File:Keno_goose.jpg. Accessed 28 October 2022.

[85] New Jersey Casino Control Commission, *TEMPORARY RULES: New Color Wagers in Roulette*. Online at http://www.state.nj.us/ casinos/actreg/reg/docs_temp_regs/t080521.diamondroulette.pdf. Accessed 9 July 2012.

[86] Nolan, William I., *The Facts of Baccarat*. Gambler's Book Club, Las Vegas, 1976.

[87] Northrop, Kate, New York Lottery draws identical Take 5 results two drawings in a row. *Lottery Post*, 28 October 0222. Online at https://www.lotterypost.com/news/343012. Accessed 28 October 2022.

[88] Player, B.J., *Here's How to Even up the Odds at Roulette*, in *The Big Winner's Systems Book*. Gambling Times, Inc., Hollywood, CA, 1981.

[89] Powerball History. Online at https://www.powerball.net/history. Accessed 14 November 2022.

[90] Prigge, Matthew J., The Policy Kings of the Sixth Ward. *Shepherd Express*, 16 November 2015. Online at https://shepherdexpress.com/ culture/milwaukee-history/policy-kings-sixth-ward/. Accessed 20 November 0222.

[91] Proctor, Richard, Poker Principles and Chance Laws, in *Play Poker To Win*, W.J. Florence, editor. Coles Publishing Company, Ltd., Toronto, 1980.

[92] Proctor, Richard A., *Chance and luck: a discussion of the laws of luck, coincidences, wagers, lotteries, and the fallacies of gambling; with notes on poker and martingales*. Longmans, Green, and Co., London, 1887.

[93] Ray, Randy, How Basic Strategy Works for Spanish 21. Online at https://www.gamblingsites.com/blog/how-basic-strategy-works-spanish-21/. Accessed 26 October 2022.

[94] Radner, Sidney H., *The Key to Roulette, Blackjack, and One-Armed Bandits*. Ottenheimer Publishers, Inc., Owings Mills, MD, 1963.

[95] Reed, C. Jill, File:Big Six wheel at the Silver Legacy Resort Casino.jpg. 23 September 2007. Online at https://commons.wikimedia.org/wiki/ File:Big_Six_wheel_at_the_Silver_Legacy_Resort_Casino.jpg. Accessed 24 November 2022.

[96] Rickyar, File: BlackjackTable.jpg. Online at https://commons. wikimedia.org/wiki/File:BlackJackGame.jpg. Accessed 4 November 2022.

[97] Roberts, Stanley *et al.* *The Gambling Times Guide to Blackjack*. Gambling Times Inc., Fort Lee, NJ, 2000.

[98] Roulette Table Layouts. Online at http://www.ildado.com/roulette_table_layout.html. Accessed 18 June 2012.

[99] Royal Roulette Pty Ltd, *Royal Roulette*. Online at http://www.royalroulette.com. Accessed 28 May 2012.

[100] Rubin, Max, *Comp City: A Guide to Free Casino Vacations*, 2nd edition. Huntington Press, Las Vegas, 2001.

[101] Savard, John, The Mersenne Twister. Published in *A Cryptographic Compendium*, 2000. Online at http://www.quadibloc.com/crypto/co4814.htm. Accessed 27 November 2022.

[102] Saxon, Jonathan, Here are the payouts and final odds for Rich Strike's win in the 2022 Kentucky Derby. *Louisville Courier-Journal*, 7 May 2022. Online at https://www.courier-journal.com/story/sports/horses/kentucky-derby/2022/05/07/kentucky-derby-payouts-2022-winner-trifecta-and-superfecta/6552222001/. Accessed 13 September 2022.

[103] Scarne, John, *Scarne on Dice*, 8th revised edition. Crown Publishers, Inc., New York, NY, 1980.

[104] Scarne, John, *Scarne's Complete Guide to Gambling*. Simon and Schuster, New York, NY, 1961.

[105] Schlesinger, Don, *Blackjack Attack*, 2nd edition. RGE Publishing, Oakland, CA, 2000.

[106] Schoenberg, Frederic Paik, *Introduction to Probability with Texas hold'em Examples*, 2nd edition. CRC Press, Boca Raton, FL, 2017.

[107] Schwartz, David G., *Big Six: A Longitudinal Micro Study*. Las Vegas, Center for Gaming Research, University Libraries, University of Nevada-Las Vegas, 2011. Online at http://gaming.unlv.edu/reports/big6_study.pdf. Accessed 6 June 2012.

[108] Schwartz, David G., *Roll the Bones: The History of Gambling*. Gotham, New York, NY, 2006.

[109] Schwartz, David G., editor, *Tales From The Pit: Casino Table Games Managers in Their Own Words*. UNLV Gaming Press, Las Vegas, 2016.

[110] Schwartz, David G., *Why Double Action Roulette Will Catch On*. Online at http://vegasseven.com/blogs/felt/2012/07/05/why-double-action-roulette-will-catch. Accessed 8 July 2012.

[111] *Scientific Betting: A Common Sense Treatise on the Application of Mathematics to Betting*. Regan Publishing Corp., Chicago, IL, 1954.

[112] Scoblete, Frank, *Break the One-Armed Bandits*. Bonus Books, Chicago, IL, 1994.

[113] Shackleford, Michael, *Big Six—Wizard of Odds*. Online at http://wizardofodds.com/games/big-six/. Accessed 6 June 2012.

[114] Shackleford, Michael, *Blackjack Rule Variations*. Online at http://wizardofodds.com/games/blackjack/rule-variations/. Accessed 5 December 2022.

[115] Shackleford, Michael, *Blackjack Switch*. Online at http://wizardofodds.com/games/blackjack/switch/. Accessed 29 July 2012.

[116] Shackleford, Michael, *Craps Side Bets*. Online at http://wizardofodds.com/games/craps/appendix/5/#firebet. Accessed 23 June 2012.

[117] Shackleford, Michael, *Craps Side Bets*. Online at https://wizardofodds.com/games/video-poker/tables/deuces-wild/. Accessed 22 October 2022.

[118] Shackleford, Michael, *Gambling 102: The Best Strategies for All Casino Games*. Huntington Press, Las Vegas, 2005.

[119] Shackleford, Michael, *Probabilities in Bingo II*. Online at http://wizardofodds.com/games/bingo/probabilities/2/. Accessed 28 October 2012.

[120] Shackleford, Michael, *Three Card Poker*. Online at http://wizardofodds.com/games/three-card-poker/. Accessed 19 July 2012.

[121] Shackleford, Michael, Two G's. Online at https://www.youtube.com/watch?v=ao3zTJa2fwM. Accessed 28 October 2022.

[122] Shampaign, Charles E., *Handbook on Percentages*. Gambler's Book Club, Las Vegas, 1976 reprint of the 1930 edition.

[123] *The Shotwell System*, in *The Big Winner's Systems Book*. Gambling Times, Inc., Hollywood, CA, 1981.

[124] Sic bo table.png. Online at http://en.wikipedia.org/wiki/File:Sic _bo_Table.png. Accessed 16 May 2013.

[125] Sklansky, David, and Alan E. Schoonmaker, *DUCY?: Exploits, Advice, and Ideas of the Renowned Strategist*. Two Plus Two Publishing LLC, Henderson, NV, 2010.

[126] Smith, Jeff, *Fred Dakota Founded Native American Casinos—In A U.P. Garage*. Traverse, Northern Michigan's Magazine, October 2009. Online at http://www.mynorth.com/My-North/October-2009/Fred-Dakota-Founded-Native-American-Casinos-In-A-UP-Garage/index.php?cparticle=1&siarticle=0#artanc. Accessed 20 July 2012.

[127] Smith, John, *No Limit: The Rise and Fall of Bob Stupak and Las Vegas' Stratosphere Tower*. Huntington Press, Las Vegas, 1997.

[128] Snyder, Arnold, Inside the Over/Under. *Blackjack Forum* **IX**, #3, September 1989, p. 5.

[129] Snyder, Arnold, The Unbalanced Over/Under. *Blackjack Forum* **XI**, #1, March 1991, pp. 7–11.

[130] Spanish Blackjack. Online at http://www.casinogaming.com/tutorials/tablegames/spanish21.html. Accessed 18 June 2012.

[131] Stuart, Lyle, *Casino Gambling for the Winner*, revised edition. Lyle Stuart Publishing, Secaucus, NJ, 1980.

[132] Stupak, Bob, *Yes, You Can Win!* Galaxy Publishing, Las Vegas, 1992.

[133] Su, Francis E. *et al.* "Perfect Shuffles." *Math Fun Facts*. Online at http://www.math.hmc.edu/funfacts. Accessed 27 June 2012.

[134] Thorp, Edward O., *Beat the Dealer: A Winning Strategy for the Game of Twenty-One*. Blaisdell Publishing Co., New York, NY, 1962.

[135] Thorp, Edward O., *Beat the Dealer: A Winning Strategy for the Game of Twenty-One*, second edition. Vintage Books, New York, NY, 1966.

[136] Thorp, Edward O., and William Walden, *A Favorable Side Bet in Nevada Baccarat*. Journal of the American Statistical Association **61**, #314, Part 1 (June 1966), pp. 313–328.

[137] Tilton, Nathaniel, *The Blackjack Life*. Huntington Press, Las Vegas, 2012.

[138] Time Anchor, File; Poker Palace1.jpg. Online at https://commons.wikimedia.org/wiki/File:Poker_Palace_1.jpg. Accessed 12 June 2022.

[139] Uston, Ken, *Ken Uston on Blackjack*. Barricade Books, Fort Lee, NJ, 1986.

[140] Vancura, Olaf, and Ken Fuchs, *Knock-Out Blackjack*. Huntington Press, Las Vegas, 1998.

[141] Vaz, Matthew, *Running the Numbers: Race, Police, and the History of Urban Gambling*. University of Chicago Press, Chicago, IL, 2020.

[142] *Video Poker FAQ: Questions and Answers.* Online at http://www.pokerheroes.com/video-poker-faq.html. Accessed 10 July 2012.

[143] Walsh, Audley V., *John Scarne Explains Why You Can't Win: A Treatise on Three Card Monte and Its Sucker Effects*. Gambler's Book Club, Las Vegas, 1972.

[144] Weiss, Neil A., *A Course in Probability*. Pearson Education Inc., Boston, MA, 2006.

[145] Werthamer, N. Richard, *Risk and Reward: The Science of Casino Blackjack*, 2nd edition. Springer, Cham, 2018.

[146] White, Jeff, *The Three Two Roulette System*. Online at http://www.silveroakcasino.com/blog/online-roulette/three-two-roulette-system.html. Accessed 16 May 2013.

[147] Winterblossom, Henry T, and Harry Houdini Collection (Library of Congress). *The Game of Draw-poker, Mathematically Illustrated: Being a Complete Treatise On the Game, Giving the Prospective Value of Each Hand Before And After the Draw, And the True Method of Discarding And Drawing, With a Thorough Analysis And Insight of the Game As Played At the Present Day by Gentlemen*. Wm. H. Murphy, New York, NY, 1875.

[148] Wong, Stanford, *Basic Blackjack*. Pi Yee Press, Las Vegas, 1995.

[149] Wong, Stanford, *Blackjack Secrets*. Pi Yee Press, Las Vegas, 1993.

[150] Wong, Stanford, Casino Mistakes—Part II, in *American Casino Guide*, 1999 edition, Steve Bourle, editor. Casino Vacations, Dania, FL, 1999, p. 213–215.

[151] Wong, Stanford, *Professional Blackjack*. Pi Yee Press, Las Vegas, 1981.

[152] Wong, Stanford, *Wong On Dice*. Pi Yee Press, Las Vegas, 2005.

[153] Zender, Bill, *Casino-ology: The Art of Managing Casino Games*. Huntington Press, Las Vegas, 2008.

[154] Zender, Bill, *Casino-ology 2: New Strategies for Managing Casino Games*. Huntington Press, Las Vegas, 2011.

Index

Printed in the United States
by Baker & Taylor Publisher Services